NUCLEAR TITANS

Edited by

Soumya Awasthi and Shrabana Barua

Vitasta

Published by
Renu Kaul Verma
Vitasta Publishing Pvt Ltd
2/15, Ansari Road, Daryaganj
New Delhi – 110 002
info@vitastapublishing.com

ISBN 978-81-960413-7-3

First Edition 2023
MRP 595

Edited by Kanagam King
Cover and layout by Somesh Kumar Mishra
Printed by Chaman Enterprises, Delhi

Dedicated to
the Atoms in the service of the nation

Thank you for making the Buddha Smile

CONTENTS

SECTION C: THINKERS/STRATEGISTS

SECTION D: DIPLOMATS

FOREWORD

I am happy to write this foreword for the book *Nuclear Titans* being brought out by two research scholars of the Centre for International Politics, Organization and Disarmament (CIPOD), Jawaharlal Nehru University, New Delhi. The initiative to understand the evolution of the unique Indian policy related to nuclear technology through biographical tales of those who directly or indirectly contributed to shaping it is indeed a novel one.

Among various instruments of national power, technology has become the most prominent, being the key driver of critical elements like economic and military strength and public opinion, among others. Soon after the emergence of technology with very high strategic potential, restrictive control regimes usually get established by those who acquired it to secure their pre-eminence and strategic advantage and prevent the emergence of a challenge to their supremacy. On the other hand, India has advocated a more harmonious, non-discriminatory, and universal world order sans coercive approaches between nation-states. This philosophy embedded deep in Indian culture has influenced the evolution of our policies regarding technologies with a robust strategic dimension. Even so, the country's security interests, from time to time, have necessitated a sound strategic culture that keeps us ahead of the curve in the fast-changing, technology-driven paradigms. Therefore, a close collaboration between people with technology foresight and strategic thinking is the need of the hour.

In the above background, the emergence of nuclear weapons was a watershed moment in the history of mankind. While India did actively participate in the international discussion related to governance in nuclear matters, their outcomes, by and large, have not been to our liking. The development of nuclear technology in India to meet civilian and military needs and related policy approaches must be seen and understood in this context. Thanks to India's philosophical roots, adherence to the principle of self-reliance, responsible approach, reasonable weightage, and response to the country's long-term interests, India did have the courage to pursue an independent path while remaining engaged with the world community. The strategy had to be continuously calibrated, considering our growing scientific strengths and being aware of our weaknesses. In my view, this approach has served the country's best interests while keeping the pathways for the future open. India could, in principle, leverage nuclear technology to rapidly address, for herself and the world, fundamental challenges before mankind. These relate to sustainable access to clean energy supply in required amounts for societies yet to meet their developmental aspirations, anxiety-free nuclear energy deployment with the near elimination of long-term radioactive waste. Even as we seek to reach a high Human Development Index worldwide, avoiding and mitigating existential threats from climate change will enhance global peace and security, the nuclear technology will become eminent. This would need a comprehensive discussion and action both domestically and internationally. Large groups of strategic thinkers working in tandem with scientific groups are thus essential.

Mobilising an India focussed comprehensive discussion that enables an independent view as envisioned above has been a challenge. Seeing growth in the community of young strategic thinkers is a very healthy sign in this context. As a nation with the largest population in the world and with the most significant demographic dividend that is the prime mover of its fast-growing economy, India is quickly gaining prominence on the world scene. Thus, India must pursue strategies to leverage this advantage and be proactive in shaping the new world order that best serves our interests. This requires insights into evolving geo-political paradigms on the one hand and strategies for leveraging our national power on the other.

Individual decisions that lead to inflexions in the trajectory are often driven either by the need to respond to a situation that has arisen or by proactive actions to derive a desired outcome. People at the decision-making level take these decisions based on their own insights, understanding, supporting information, and available research. This part needs mature strategic analysts. This book is a good beginning for young strategic analysts to understand the evolution of Indian nuclear policy through biographical tales related to some of the significant people in the nuclear domain.

The connections between events as they happen and their strategic underpinning are not always straightforward and obvious. For example, Bhabha decided to set up Tarapur Atomic Power Plant on a turn-key basis through a global tender. And around the same time, he decided to develop a uranium mine at Jaduguda through indigenous efforts, even when it was clear that the ore grade was far below what one would consider economically viable. The decision on Tarapur itself was made with a complete understanding that the workhorse technology for India's self-reliant nuclear power programme could be different, which turned out to be PHWR (Pressurized Heavy-Water Reactor). It is also worth noting that the first pair of PHWRs at Rawatbhata, Rajasthan, was set up as a collaborative effort, not through a vendor-buyer transaction, as in Tarapur. Though a contradiction within themselves, each of these decisions was crucial for the success of our self-reliant nuclear programme. Such nuances must be understood to appreciate policy formulation and implementation at relevant times. As we advance, with the discussion becoming more profound and comprehensive, with the maturing of the community and increasing engagement of young researchers, I am confident the quality of narration and analysis will become more refined.

Technology is a tool for the augmentation of human capability. Whether that capability is leveraged for more extensive human welfare or for exploitation driven by animal instinct which we all naturally possess, is a matter of value system that we subscribe to, individually and collectively. As India rises globally, we have much to offer towards shaping the world as a more enlightened human habitat free from cliff-edge catastrophes. Realising this is a matter well beyond the technology people alone. The community of social scientists have an important role here. This book is a welcome step in that context.

I want to thank all contributors for bringing out important aspects from the life stories of scientists, statesmen, strategists, and diplomats responsible for shaping policies related to nuclear technology in India in its formative years. Dr Soumya Awasthi and Dr Shrabana Barua deserve special compliments for compiling the book.

Anil Kakodkar
(Chancellor, Homi Bhabha National Institute
Chairman, Rajiv Gandhi Science & Technology Commission
Former Chairman, Atomic Energy Commission)

PREFACE

It was in 2015 that we began our association with the Jawaharlal Nehru University, New Delhi. We joined the Centre for International Politics, Organization and Disarmament (CIPOD) as Research Scholars. Today, as we write, the name JNU unfortunately triggers strong sentiments in the minds of readers. The fact that the book has been edited by two JNUites of the Kanhaiya Kumar period of branded 'anti-nationalism' or '*tudke tukde gang*' might have created an unfavourable impression on some causing them to immediately put it down, though we hope you are not prey to such narrow-mindedness. Some may, in turn, want to know what it is that makes JNU stand apart. After all, more recently, JNU alumni have been winning accolades, including Geetanjali Shree, the first Hindi author to win the International Booker Prize!

Notwithstanding the controversies, in addition to the immeasurable contributions made by the various departments at JNU, it is noteworthy to underline that the School of

International Studies has produced some of the finest intellectuals, policy makers and diplomats who have shaped foreign relations of India, its nuclear policies, its security conduct, and guided many other such matters considered 'mainstream' and crucial for a country's progress in the twenty-first century.

But rest assured, this volume is definitely not an attempt to vindicate the status of JNU, a university that stands for diversity and debate, for inclusiveness and knowledge. Instead, it hopes to showcase how people, irrespective of their background, class, status, field of knowledge, but with the right amount of zeal, vision, grit and hard work, can, and indeed do, make a difference in the moulding of a modern nation state. It is also a work that reminds all of us that each successful person is more than what his/her career shows and that each accolade a country achieves has many hands that work on it. In this volume, the topic that we focus on is the nuclear path India has taken for itself, made possible by the efforts of its people.

Though the field of social science research has drawn our individual interests since our post-graduation days at the University of Delhi in 2010-11, we found ourselves intrigued by the writings on nuclear history and strategic thought, introduced to us during our PhD course work period. As scholars of the Diplomacy and Disarmament Programme, we began reading about security studies, about the minimalist and maximalist debate on deterrence propounded by Blackett and Wohlstetter, about the nuclear history of India, about what it means to use tactical weapons vis-à-vis a strategic one, how international security evolved from a traditional state-centric ontological subject to recognize the threats that are unconventional and multi-pronged. The point being, we were being prepared for practically understanding and realizing the implications of strategic and nuclear politics of the international system in general, and India in particular.

As much as the war stories and theoretical understanding of nuclear behaviour of states sparked an interest in us, reinvigorated by the Russia-Ukraine War in 2022, the novel Covid-19 pandemic that gripped the world in 2020 made the study of non-traditional security issues more pertinent than ever before. It is during such times that we as people truly understand the meaning of hope, the value of research and the urge to do more from what we have. With this lesson, and the larger need to bring out the stories of many people who have helped contribute and shaped the field of nuclear development in India, we bring forth this edited volume titled *Nuclear Titans.*

The core idea of researching on the life of personalities who contributed to the making of nuclear India was sowed in our minds by our professors at CIPOD in JNU, especially Prof Amitabh Mattoo and Dr Happymon Jacob, as early as 2015. As part of the paper 'India and Disarmament', comprehensive and analytical works were prescribed to us for understanding India's nuclear story – George Perkovich's master piece, *India's Nuclear Bomb: The Impact on Global Proliferation* (1999), Raj Chengappa's *Weapons of Peace: Secret Story of India's Quest to be a Nuclear Power* (2000), Itty Abraham's article *India's*

'Strategic Enclave': Civilian Scientists and Military Technologies (1992) that described how the 'strategic enclave' operated in nuclear policy making, Robert Anderson's *Nucleus and Nation: Scientists, International Network, and Power in India* to name a few.

Yet, what excited us was not to just read these commonly available sources, but the fact that we were asked to go fetch stories, unheard of, from the lives of the personality we chose to research upon as part of our assessment process. This tickled our enthusiasm as new researchers. We were encouraged to meet the person or his/her family, friends and colleagues on whom we would write our biographical assignment. We were asked to seek out stories from the secondary data that might reveal interesting information on how a certain decision was made or how nuclear politics panned out in India. The idea was to realize that the research required going beyond what is already given and adding information about a person's life that the common public or even experts on nuclear history may not know. The early life stories of the many people who played or continue to play essential roles in India's nuclear journey, as well as their motivations for doing what they did, their career paths, their success and failures, all had to be documented. This was indeed the starting point of our work, that we then decided to convert into a book, much later in 2018.

We reached out to many people with the list of personalities we had in mind. Fortunately, most of them have contributed to this volume, for which we are deeply grateful. Among them, we found special encouragement from Dr Manpreet Sethi, who showed great trust in us, despite delay in our publication process. She gave us the necessary push wherever needed. Our heartfelt thanks to Dr Anil Kakodkar, who has kindly written the foreword for us.

There are many auto/biographies of famous people, including some exclusively in the nuclear field. There are also scattered articles about the life stories of many personalities included in this volume, mostly written on anniversarial occasions and especially (though unfortunately) as parts of posthumous obituaries. Moreover, either they are read by the people in the field, the contemporaries and the disciples, or readers interested in nuclear science alone, for such a field is thought of by a layman as a subject of high intellectual fodder or even borderline boring. Though the prime focus of this book is not to draw the attention of such a readership alone through the narration of interesting and personal stories and many pictorial references even collected from primary sources, the need to bring out a volume that allows us to go through the journey of nuclear India from the perspective of its many contributors, by peeking into their lives, itself is a tempting and calling matter. Therefore, the fundamental purpose of this volume is to bring forth the untold stories from the lives of a selected few people, though many may question our criteria for the selection. The effort is to place their lives in the journey of nuclear India, to not only understand how we reached where we are today as a modern nation-state, but also gauge the mistakes to learn from them for the future.

Lastly, this volume indeed makes it possible to understand

that no matter which part of India one is from, what economic status one belongs to, what subject and interest one keeps, it is never impossible to contribute to the making of the modern democratic nation that is India.

Shrabana Barua
Soumya Awasthi

INTRODUCTION
Start of the Atomic Age and India's Nuclear Rise

Shrabana Barua

With the use of the atom bomb in the mid-1940s, power dynamics began to be determined by the absence or presence of nuclear weapons with a state. The five nuclear weapon states (NWS) – China, France, Russia (erstwhile USSR), UK and USA – leveraged their position in not just shaping the nuclear regime since the 1960s, but asserted influence in multiple forums through their acquired power, emanating from the possession of nuclear weapons and nuclear technology. While most of the world acquiesced to the nuclear order (with or without pressure), India's position on nuclear weapons was rather complicated. But before we begin with that story, let us understand what a nuclear weapon or technology really is.

The basics of nuclear science

The field of nuclear science and nuclear strategy has a fairly recent history. While the latter is a product of the making of nuclear weapons, a phenomenon that started in the early 1940s, the former can be traced back to the discovery of Uranium in 1789 by Martin Klaproth (though the mineral, pitchblende, could be found in silver mines). Because uranium is radioactive, it fits well to look back to the late eighteenth century. However, a separate subject of nuclear physics, distinct from atomic physics, emerged only in the late nineteenth century. Before that, for long, it was assumed that an atom is indestructible, as explained by John Dalton's theory in 1803. That an atom had electrons (negatively charged particles), was theorized by JJ Thomson in 1897.

Meanwhile, in Paris, in 1896, Henri Becquerel discovered that uranium is radioactive, i.e. it has an unstable nucleus or that there is a nuclear decay. At the time, neither a nucleus (consisting of positively charged protons and neutrons, which have no charge) nor a neutron was known of. Marie Curie, Pierre Curie and a few others continued to work on radioactivity. By the end of that century, three types of radiation – alpha, beta and gamma – were found to release energy. Of course, nuclear energy today is known to have great potential for electrification and other uses, being rooted as a clean renewable energy source. But back then, discoveries were still being made about the basics in the field of nuclear science.

It was in 1911 that Ernest Rutherford discovered the nucleus. The structure of the atom remained a subject of study. But nuclear physics had not yet advanced enough to provide assistance during the First World War that broke out in 1914. Hence, war strategies remained confined to traditional deployment, command and control of the army, navy and air forces and their various weapon systems (which falls under the subject of science, of course). During that time, in 1920, Arthur Eddington hypothesized about nuclear fusion taking place in the stars. A fusion process occurs when two or more nuclei merge to form a single nucleus, releasing energy as a result. It was found that two atoms of hydrogen fuse together to form one atom of helium, a process that forms the power source of the sun. Much later, in the 1950s, this process was put to use to develop the first nuclear fusion device by USA, a thermo-nuclear bomb or a hydrogen bomb.[1] Even today, scientists are attempting to find ways to harness energy from nuclear fusion, which could lead to an unlimited source of clean energy. But the atom bombs dropped in Hiroshima and Nagasaki in August 1945 were fission devices.

A fission process (the name draws from biology) is the splitting of a nucleus into two (or even three), leading to release of gamma rays or energy. While tensions were rising in Europe, German scientists Otto Hahn and Fritz Strassmann, along with Sweden's Lise Meitner, discovered nuclear fission of heavy elements in 1938.[2]

For fission to take place, a neutron is required to initiate a chain reaction of a radioactive fuel, in this case uranium-235 (isotope). Already, in 1932, James Chadwick had discovered

1 Both hydrogen bomb and atom bomb are nuclear bombs, former as a result of nuclear fusion and latter that of nuclear fission. Hydrogen bomb is a second-generation nuclear device that requires the energy produced from nuclear fission in its first stage.

2 Read Online History of Nuclear Energy, November 2020.

the neutron. Now, attempts were being made to find out how U-235 (rare fissile material) can be obtained from U-238 (naturally occurring) through 'enrichment' and how more neutrons could be produced as a by-product of a nuclear fission, to trigger the chain reaction.[3]

By 1940, it was discovered that another fissile material, named 'element 94' or Plutonium (credit for this name is given to Glenn Seaborg from the Berkeley team in 1941, and the research team in Cambridge, Britain, who also came up with the same name) could be produced from U-238. This meant there could be more fissile material made, when need be. It also meant that it was now possible to induce nuclear fission or decay, apart from natural decay in radioactive substances. In short, a nuclear bomb was possible. So was nuclear technology for civil use, by controlling the chain reaction. The power plants and technology used in current times are powered by the fission processes. As these discoveries and research make it clear, the advent of the nuclear bomb really began from the 1930s. Thereafter, the need arose for nuclear strategy.

Rise of nuclear powers and nuclear regime

Many geopolitical and geo-strategic developments emanated from Germany in the late 1930s. Since Germany designed Project Uranverein to make the atom bomb in 1938, USA

3 A 'critical mass' of uranium is required for a chain reaction to take place. Also, it was found that a slow-moving neutron better initiates a fission process than a fast one. Therefore, the idea of using a 'moderator' was introduced by Leo Szilard and Enrico Fermi.

established the famously known Manhattan Project in the early 1940s. The classical idea of security dilemma can be used to explain the creation of the atom bomb by the US. Given that World War II had begun in 1939, it was believed that whoever developed the atom bomb first, would win the war. As history shows, the first to use it, indeed did.

In USSR, the Academy of Science set up a project in June 1940, among other things, to find out about the uranium deposits in the Central Asian region. However, German invasion of Russia a year later, pushed the direction of nuclear research towards military applications. Many opined that Germany's military objectives of nuclear weapons were rather impractical as it required immense resources. Soon they shifted their focus to rockets. In Britain too, work on nuclear physics was on. This was the time when a few people from India were involved in work on nuclear reactors and physics; their stories can be found in the following chapters of this book. Meanwhile, the US government sanctioned projects that hosted scientists from all around, including Britain, Germany, Russia, in an effort to create the atom bomb. This – the Manhattan Project – was brought under military command. By 1942, the first controlled nuclear chain reaction was made possible under the leadership of Enrico Fermi. With Robert Oppenheimer as Director, a secret project was on at Los Alamos,[4] to create the atom bomb. In the desert of New Mexico, at Alamogordo, the first nuclear

4 There were also simultaneous projects going on in the US, with the assistance of UK and Canada.

test was conducted successfully on 16 July 1945. Oppenheimer's words 'now I become Death, destroyer of worlds', announced the start of the atomic age, an age of power through the means of the atom bomb.

USSR had also begun a modest nuclear research programme in 1942. But it is only the use of the atom bomb, which subsequently ended the war, that gave steam to Russia's aim of detonating a nuclear device. This, it did in August 1949. It is said that between 1949 and 1989, as many as 456 atomic and thermo-nuclear devices were tested at Semipalatinsk Test Site (STS) (CTBTO, 2022). In 1952, UK concluded its first test, declaring itself a nuclear power. In 1960, France became the fourth state to test nuclear weapons. Like USSR, France too began expansive testing. From 1966, it is said to have tested forty-one devices till 1974 (Cho, 2021). All of them are thought to be above ground tests, conducted in French Polynesia. It is in 2013 that a report, based on declassified material, brought to light that ninety per cent of the people were exposed to radioactive fallout as a result of the French testing during the time in context (Ibid.).

No doubt, there was now a nuclear threat that had created the need for formulating nuclear strategy and a global nuclear regime. That the very first resolution of the UN General Assembly gave birth to the United Nations Atomic Energy Commission (UNAEC) in January 1946, speaks of the importance of dealing with nuclear powers. The same year, the US government floated the Baruch Plan. The significance of it was not so much its objective of controlling the use of the atom for peaceful purposes while encouraging nuclear technology, as much as its proposal to eliminate all atomic weapons (at that time only under possession of the US) and other weapons of mass destruction. This is seen as a lost chance in history, as USSR rejected this plan, fearing American monopoly of nuclear matters. India, represented by Vijaya Lakshmi Pandit at the UN discussions over this plan, supported the idea of using nuclear capability for peaceful purpose alone, but was not convinced about US's plan. What the withdrawal of the Baruch plan did was start an arms race in the backdrop of a Cold War.

The UN, as a world organization, attempted to fulfil its objective to 'save succeeding generations from the scourge of war', by taking stock of the nuclear situation. At the General Assembly in 1952, a resolution led to the creation of the UN Disarmament Commission (UNDC) under the Security Council command. Disarmament thereon became an important agenda in the security narrative, especially of nuclear weapons. To be noted is the fact that this attempt took varying dimensions. For instance, the Ten-Nation Disarmament Committee was formed in 1960, which became the Eighteen-Nation Disarmament Committee in 1969, eventually culminating into the Conference on Disarmament (COD) in 1978.

Interestingly, beyond disarmament, the nuclear narrative was taking a different turn, at the behest of the US. The atom was being portrayed as a positive harbinger of change for the future. In 1953, for instance, US President D Eisenhower gave his famous 'Atoms for Peace' speech. It was an attempt to pacify the publics who were worried about the horrors of the atomic world as well as to give hope of building a transparent and efficient nuclear world, where use of the atom could change life for the good. By then, nuclear weapons were being considered as weapons of deterrence, something that was axiomatically to avoid world war three. The International Atomic Energy Agency (IAEA) was set up in 1957 as an autonomous body under the UN aegis. The world was moving towards devising ways to control and cooperate over nuclear issues. Such efforts were also substantiated by the efforts at non-proliferation of nuclear weapons.

In 1963, the Non-Proliferation of Nuclear Weapons Treaty (NPT) was formulated with three pillars at its core – non-proliferation, disarmament and peaceful use of the atomic/nuclear energy. At the same time, the Partial Test Ban Treaty (PTBT) was being signed by USA, USSR and UK to abstain from testing nuclear devices in the atmosphere, in outer space and under water (only underground was not prohibited).[5]

In fact, it was India that proposed a ban on nuclear testing through a Standstill Agreement as early as 1954. However, as we now understand, India's early negotiations at restricting the ills in the atomic age did not go as envisioned. Of course, that India's vision itself was marked by ambiguity is often highlighted. When the NPT was put to signature in 1968,

5 France and other states continued to test over ground however.

India was among the four states (including Israel, North Korea and Pakistan) that refused to sign the international document. One of the reasons for this was China. China had become a nuclear state in 1964, testing a fission device at the Lop Nur site of the Gobi Desert. Within three years, it had tested a hydrogen bomb as well. India felt insecure, especially after its debacle during the 1962 India-China war. But more significantly, India's reservations with the NPT were because of the fact that as per the new regime-in-the-making, only the five states that had tested nuclear devices were considered as Nuclear Weapon States (NWS). India considered the NPT, which was put into force in 1970, as an unfair deal, a treaty that divided the world into nuclear haves (NWS) and have-nots (Non-NWS). It seemed to permit vertical proliferation while prohibiting horizontal proliferation.

India stood its ground and refused to give in to what it believed was a discriminatory international nuclear regime, whether it be the NPT or later the Comprehensive Test Ban Treaty (CTBT that replaced the PTBT in 1996). That being so, how has India become an exceptional case today?[6] Exceptional, because despite conducting nuclear tests and not acquiescing to international rules it finds unfair, in 2005, it managed to negotiate a civil nuclear deal with the leading rule maker of the West, USA. On the other hand, it seeks to become a part of the Nuclear Suppliers Group (NSG), initially created as a response

6 C Raja Mohan calls India a case of nuclear 'exceptionalism' in this book *Crossing the Rubicon: The Shaping of India's New Foreign Policy.*

to India's nuclear test of 1974, while continuing to remain a responsible nuclear nation, standing on a moral high ground. What does India's nuclear policy and nuclear diplomacy therefore look like? And more importantly, as is the intent of this book to show, who shaped these pathways for India?

India's nuclear green shoots of exceptionalism

This volume is not a validation of India's stand on the nuclear regime. Nor is it about trying to understand this trajectory and help assess the future. It is simply the stories of people who toiled to make India the exceptional nuclear case that it is today. What that case is, is outlined in this section. Any lessons that help a reader envision the future is a by-product of this effort.

At the time when the First World War began, in India, the Indian Science Congress Association (ISCA) was being founded in Kolkata with the goal of encouraging research on science and application of technology. There were other scattered attempts at scientific institution building. Soon its fruits became visible. Scientists like HJ Bhabha and Meghnad Saha (who is known to have had disagreements with Bhabha's way of functioning at the AEC and otherwise) were among prominent Presidents of the ISCA. While Kolkata became one of the early centres of scientific valour in India, Bangalore and Bombay soon caught up. Already, as early as 1909, the Indian Institute of Science (IISc) was set up in Bangalore with support from Jamsetji Tata. In 1934, CV Raman (who also strengthened the Department of Physics at IISc and worked closely with Bhabha, and later Vikram Sarabhai, among others) founded the Indian Academy of Science (IAS) in the same city, with the objective of 'promoting the progress and upholding the cause of science' (IAS website). Yet, it is the scientific vision for India, led by the first Prime Minister, Jawaharlal Nehru, as well as the socio-economic requirements of the time, that culminated into efforts at national planning for development, including in the field of nuclear science. At this stage, the ideas of those such as Meghnad Saha (who set up the Institute of Nuclear Physics in Kolkata and founded the journal *Science and Culture*), Shanti Swarup Bhatnagar (who led the Council of Scientific and Industrial Research, CSIR), Homi Bhabha and many others were critical.[7]

As the atomic age unfolded in the 1940s, India was emerging as a newly independent nation-state, grappling with various socio-economic problems. Nuclear issues were not a priority therefore. But Jawaharlal Nehru did not fail to acknowledge the usefulness of nuclear power for the development of a modern state, while making his abhorrence towards nuclear weapons vocal. He said, 'the future belongs to science and to those who make friends with science and seek its help for the advancement of humanity' (*Selected Works of Jawaharlal Nehru, 8: 806-8*). By this time, among the many Indian scientists who had been studying nuclear science at home and abroad, Homi Jehangir Bhabha understandably stood out. In Chapter 1 of

7 see Robert Anderson's *Nucleus and Nation*, Chapter 6 and 7.

this book, titled 'Homi Bhabha: The Father of Nuclear India', Rajarshree Chowdhury captures the life and works of such a tall personality, as comprehensively as possible within the limits of one academic piece. Bhabha (popularly called the Father of India's Nuclear Programme) was close to Nehru at a personal level and worked towards shaping India's nuclear strides even before its independence. This is not to give all credit to Bhabha himself. In fact, the very attempt is to know more and go beyond the obvious. Yet, Bhabha's relentless efforts for establishing a nuclear programme cannot be matched by others. As a result of that effort, the Tata Institute of Fundamental Research (TIFR) was established in 1945. This was facilitated by a fund Bhabha managed to obtain (with an estimated budget of Rs 80,000) from JRD Tata (Perkovich, 1999:16).

When the world witnessed the terror of the atom bomb in Japan, India under Nehru was best described as representing 'moral absolutism'. The atom bomb was the horror that needed to be eliminated. In 1946, at the time when the US established the Atomic Energy Commission (AEC), transferring atomic energy as a subject to civilian control, India took note (though India's AEC was based on the British model). The same year witnessed the establishment of the Atomic Energy Research Commission (AERC) in India. In 1948, India's AEC was established, featured with extreme secrecy. Both AERC and AEC were under the Department of Scientific Research, and then became attached to the Department of Atomic Energy (DAE), established in 1954. In the 1950s, therefore, India pushed for disarmament and non-proliferation at a global stage and strived to gain out of the peaceful use of the atom. This it did by taking various steps. In 1951, for instance, India signed a nuclear cooperation agreement with France. The very next year, a four-year plan was announced that sought to harness atomic material in the country and develop India's nuclear capability. Of this, monazite, as easily available, was to be to processed to obtain thorium. Soon enough, in 1953, India looked at exporting thorium nitrate, especially to China. However, the USA intervened as per its laws under the Mutual Defense Assistance Act and agreed to buy the same from India in the future. This episode made it clear that though India hoped to be a flagbearer of disarmament and non-proliferation, it was in no way ready to be dictated by external variables that impinged on its national interest and sovereignty – a sign of India's exceptionalism in the making.

In 1954, Bhabha presented a three-stage nuclear programme for India. It included the following steps:

- U-238 to be used in reactors as fuel to produce power. P-239 will be left as a by-product of this process.
- P-239 will be used, along with Th-232 (which is readily available in the country because of monazite reserves) as fuel to initiate a fission process. A by-product of U-233 will be obtained.
- Breeder reactors will be constructed, where burning of U-233 (and Thorium) will in turn produce more U-233 than it consumes. This will provide an unlimited supply of fuel therefore.

There were of course apprehensions and even opposition, about the feasibility of a three-stage programme, about the increasing secrecy of the plans by Bhabha and of the overbearing budget that he demanded for the projects in Bombay. Luckily, in 1956, the Apsara research reactor achieved criticality and dampened boisterous criticism. Here, scientists including Homi Sethna, KS Singhvi, Raja Ramanna, AS Rao and others need to be celebrated. This was after all a defining moment in India's nuclear history. At the time, in 1955, even Canada offered to build the 40MW CIRUS reactor (in 1963 Canada agreed to also build the Rajasthan Atomic Power Station or RAPS-1). These progresses led to the Indian government officially accepting the three-stage nuclear programme of India in 1958.

At the level of nuclear policy too, Bhabha had garnered increasing authority. In addition to all the posts he held, he was made Director of the Atomic Energy Establishment (AEE, later renamed BARC), set up in 1957 at Trombay. Bhabha also went ahead to have the Phoenix nuclear plant built in 1961 (for plutonium-uranium extraction).

It was very clear from the beginning that India's nuclear control would not be in the hands of the military. Instead, it would have a rather democratic process (with the Prime Minister at its helm). At this stage, India's desire to remain afloat in an increasingly tightening nuclear regime needs to be underlined. The IAEA was entrusted with the powers of inspection and safeguard of nuclear matters. India (along with others such as USSR) voiced strong opinions against IAEA's control of plutonium stocks of a country, and indeed managed to press its point. Similarly, India's views on IAEA's inspection over enriched uranium prevailed (see Weiss, 2010). As noted above, India was simultaneously involved in disarmament and non-proliferation discussions at the international level. In fact, in 1964, it laid down a proposal on nuclear non-proliferation. However, the next decade unfolded into a period of continuity with some piecemeal yet monumental changes. As is known, India conducted its first nuclear tests at Pokhran in 1974. Why and how did India begin to change its stance in the nuclear narrative?

The traditional answer to this question has been – due to security and strategic reasons around India's neighbourhood. The war with China in 1962 and China's nuclear tests in 1964 stand highlighted in this context. On the other hand, what is less discussed is India's aim to become self-sufficient at the nuclear level and its desire to chalk out an independent path for itself. Among others, Raja Ramanna's role appears pivotal in this case, which has been discussed in depth in Chapter 3, titled 'Raja Ramanna: The Nuclear Physicist with a Piano', by this author. Dr Ramanna, who was inducted into the TIFR in 1949 (where there was a separate room allotted for him to practice music), had noted that the AEC was founded not just with the aim to develop nuclear power but with the aim of making India self-sufficient. The CIRUS reactor came in as a part of this effort, backed by the operation of the Phoenix plant.

Lal Bahadur Shastri's entry at the nuclear helm after

Nehru's death made things difficult. Shastri, a Gandhian by practice, did not give in to the ideas of the 'nuclear enclave'. However, with time (especially after his visit to Trombay in 1965), he allowed the dice to get rolling with the SNEPP (Study of Nuclear Explosions for Peaceful Purposes). Raja Ramanna was made to head this.

The unfortunate demise of both Shastri and Bhabha in 1966 also marked a shift in the nuclear story. The new Prime Minister, Indira Gandhi, described by Sitakanta Mishra as 'The First Among Pioneers' in Chapter 5, was a rather determined personality. It is under her political leadership that India conducted its first nuclear tests in 1974. At the same time, Dr Vikram Sarabhai was left with the huge task of taking over some of Bhabha's projects, especially as the new head of the AEC and the DAE. As it is, Bhabha had propelled Sarabhai to lead from the front of the nuclear programme in process. Yet, it is in the field of space research and institution building that Sarabhai outshines others. The multitude, vibrancy and dynamics of his work have been very succinctly discussed by Sameer Patil and Arun Vishwanathan in Chapter 2, 'Vikram Sarabhai: An Institution Builder'. Remarkably, Sarabhai helped establish the Indian National Committee for Space Research in 1962, renamed Indian Space Research Organisation (ISRO) in 1969, under the DAE.

There were yet others such as Homi Sethna and MGK Menon who filled in for Bhabha at the AEE/BARC and TIFR respectively. The new leadership, especially of Indira Gandhi and Sarabhai, marked a departure from the previous one. The

SNEPP was not seen as a priority. Some disagreed on various fronts, often leading to internal debates within the DAE among the new 'enclave', including engineers and scientists.

The NPT, which came into force in 1970, was seen as one reason for Indira Gandhi's change in attitude towards allowing nuclear tests in May 1974. Another reason was the war with Pakistan in 1971. By then, Sarabhai had died and Homi Sethna was leading Operation Smiling Buddha. New Delhi declared the tests as only Peaceful Nuclear Explosions (PNE). This nomenclature did not fend off harsh reactions by international nuclear watchdogs. In particular, the Nuclear Suppliers Group (NSG) was created in 1975 to prevent the transfer of nuclear technology, material and equipment to NNWS. This export control regime continued to evolve over time. Today, India still remains outside of it, though it seeks its membership.

In the years preceding Operation Smiling Buddha, a debate had begun amongst strategy experts in particular on how India must proceed with the nuclear question. The political dispensation listened carefully. Unlike in the past, the secrecy surrounding nuclear issues was waning. George Perkovich notes that during 1970-71, three camps emerged over nuclear issues and foreign policy. One, which opposed nuclear weapons; one that advocated a crash programme for building a small nuclear arsenal with delivery system; and one that proposed developing a long-term steady nuclear capability (1999:156). Among these, K Subrahmanyam, who headed the Institute for Defence Studies and Analyses (now Manohar Parrikar - IDSA), backed the third position. This is in fact the position that India ended up taking, i.e. a gradual process of making India nuclear capable in terms of both deployment and infrastructure. Similarly, he was an ardent supporter of the No-First-Use (NFU) policy, that continues to be a part of India's nuclear policy, amidst growing voices that apprehend a change in this by the current Bharatiya Janata Party (BJP) led government. In any case, Subrahmanyam's contributions to India's nuclear strategy and policy, both civil and military, are quite substantial. Chapter 7 titled 'K Subrahmanyam: The Scholarly Strategist' written by Soumya Awasthi delves into his life, ideas and lessons collected.

In the decade succeeding Operation Smiling Buddha, India was a more self-aware and established nation. It had emerged as a leader of the Non-Aligned Movement (NAM) and stuck by its decisions at the international nuclear regime. On the other hand, work was also going on in the verticals of space and rocket research, as well as improving India's defence delivery system and strategy. While Sarabhai was the forerunner of the space programme, it was APJ Abdul Kalam who pushed ahead India's rocket programme, operating from Thumba Equatorial Rocket Launching Station. By 1968, Satellite Launch Vehicle (SLV) development studies were undertaken at ISRO. Even before that, the Special Weapons Development Team (SWDT) had begun work on missile development at the Defence Research and Development Organisation (DRDO). Kalam's shift from ISRO to DRDO eventually gave momentum to the Integrated Guided Missile Development Programme (IGMDP) in the 1980s. Chapter 4, written by Rajaram Nagappa brings out the story of the 'Missile Man of India'.

Soon after India tested its first surface-to-surface missile (Prithvi) in 1988 and the indigenous inter-continental ballistic missile (Agni) in 1989, international export control regime came in, in the form of the Missile Technology Control Regime (MTCR), established by the G-7 states. India resisted, just as it continued to resist the NPT that was extended indefinitely in 1995. Similarly, in 1996, the Comprehensive Test Ban Treaty replaced the PTBT (Partial Test Ban Treaty). Though India helped in formulating an international regime that restricted nuclear tests, and also remained a votary of disarmament, it never signed any of these treaties. Yet, it continued to justify its stand while advocating disarmament, becoming quite an outlier at the time. For instance, the Rajiv Gandhi Action Plan was proposed at the special session on disarmament at the United Nations General Assembly (UNGA) in 1988. It sought to rid the world of nuclear weapons in two decades. It was never accepted internationally.

Arundhati Ghose represented India in the Conference on Disarmament in the mid-1990s. As a diplomat, she noted the discrepancies in the resolutions passed by the UNGA in 1996 and the one in 1995, where India had participated. She stated that there was no 'reasonable and negotiated finite span of time' within which the NWS were to eliminate their nuclear weapons. This is why the treaty was seen as unequal and discriminatory. Vijaylaxmi Yadav indulges in bringing out the efforts made by 'Arundhati Ghose: The Proud Indian at the NPT Negotiations' in Chapter 10. The line of arguments given by Ghose continued to be followed by Indian representatives in coming years.

Revealingly, the narrative in India regarding nuclear science and technology had undergone a change from the 1950s to the 1970s and further to the 1990s. In 1994, a detailed survey conducted to gauge public opinion on nuclear issues showed that Indians, though more concerned by issues of communalism, economic conditions, terrorism, etcetera, nevertheless voted that India must have the nuclear option available.[8] At the same time, disarmament was also supported. Thus, the phase of India's nuclear ambiguity visibly still lurked on.

Between 1974 and 1998, there were evidences of dispersed attempts at nuclear testing. The 1995 episode, when Narasimha Rao as Prime Minister gave the green signal for nuclear testing (though not of the nature that scientists demanded), is crucial to understand the transition that India was making. The detection of the same by American intelligence led to tightening of the noose on India. While the US appreciated the civilian control of nuclear matters in India, especially during a visit by Defence Secretary William Perry in 1995, India was urged to adopt arms control measures for stability in the region (Perkovich, 1999:355).

It is essential to also note the growing civilian and military voices that fed into the democratic decision making of India's

8 See Report on 'A Study of India's Nuclear Choices' by Marketing and Research Group at the John B Kroc Institute for International Peace Studies, University of Notre Dame, 1994-5 (as quoted in Chapter 13 of Perkovich's *India's Nuclear Bomb*, 1999).

nuclear story. From academic institutions and policy makers to public opinions over nuclear matters, there was a vibrant debate on the subject, unlike the utter secrecy that marred the nuclear programme at its initial stages. For instance, Air Force Commodore Jasjit Singh opined that because India did not have an integrated military and strategic planning structure in place, it took away from India's ability to do better in nuclear matters. In Chapter 8, Manpreet Sethi brings out the views and contributions of 'The Nuclear Strategist with the Red Turban', highlighting Singh's works and writings on importance of nuclear energy, NFU, and India's security needs emanating from situations in the neighbourhood. It is perspectives such as his that give readers a diverse understanding of India's nuclear story. A peep into the life and times of such persons adds invaluably to one's understanding.

Meanwhile, India had made good progress with its nuclear programme. The period of 1980s and 90s are ripe with examples. In November 1982, the Tarapur fuel reprocessing power reactor (under IAEA control) was commissioned. Reprocessing of spent fuel is an important aspect of India's aim to function as an independent state with indigenous nuclear capability. By 1986, India was reprocessing spent fuel at Tarapur from the Madras Atomic Power Station (MAPS). In 1985, the 100 MW Dhruva reactor went critical, producing weapons-grade plutonium. In 1989, the 220 MW pressurised heavy water reactor was operational at Narora-1. The second unit became operational in 1992, the same year when the Kakrapar reactor also was made operational. In 1996, the 30

KW KAMINI (Kalpakkam Mini) research reactor attained criticality.

1998, of course, was a watershed year, with India conducting Operation Shakti in Pokhran. This time, however, led by Prime Minister Atal Bihari Vajpayee, India went overtly nuclear. Vajpayee announced that India tested a fission device, a low-yield device, as well as a thermonuclear device, i.e. a fusion device. Two days after, on 13 May, two more tests were conducted. India was swift to simultaneously propagate its policy of NFU and credible minimum deterrence. As expected, sanctions came pouring in.

In this book, the story of nuclear India from 1998 onwards has been analysed through the perspectives of two personalities. First, in Chapter 6, Yasemin Develi brings out the motivations, political will and factors leading to Atal Bihari Vajpayee's decision to making India openly nuclear. No doubt such a feat in Indian history is the workmanship of many. Some of those people find place in the various anecdotes within the chapter. However, the focus of the chapter remains the ideological factors of the BJP-led government and Vajpayee's personal understanding of the importance of the security situation in India, given his earlier experience as the External Affairs Minister in the Morarji Desai cabinet.

Secondly, Renjie Feng adds value by delving into the contemporary socio-political space of nuclear matters by penning a chapter on 'C Raja Mohan: An IR Scholar with a Nuclear Physics Backgroud'. This inclusion helps understand the more recent debates and policies in India's nuclear story.

However, including a personality who is very much active in his professional life is both a boon and a bane. By peeking into Raja Mohan's work, especially as he tries to explain how India managed to 'cross the rubicon'[9] in 1998, as well as managed to emerge as a case of 'nuclear exceptionalism' by signing the Indo–US Civil Nuclear Deal in 2005, elevating India's position internationally, we tread a very dynamic and pertinent territory. India's engagement with the US on nuclear matters have been dealt with pragmatically by Raja Mohan and many others. That nuclear weapons continued to be viewed as a symbol of power was quite evident. Nevertheless, there are many more scholars, strategists, and journalists who have arguably done enough work on the subject of India's nuclear story and deserve to find space in this volume. Dedicated to those, there need to be many more volumes on the subject, because it is not possible to work on such a topic without parsimonious selection.

In the aftermath of India's exceptional status, which itself was validated by the deal with US, we find various elements in India's nuclear trajectory that need attention. Of them, India's official entry into the MTCR regime in 2016 is an important episode. On the basis of its changed status, India also continues its attempt to enter the NSG, to enable trade in nuclear technology. In the field of nuclear energy, India aims to shift to renewable energy sources, in line with the Sustainable Development Goals, and has accordingly invested in building

9 Reference to C Raja Mohan's book, *Crossing the Rubicon: The Shaping of India's New Foreign Policy* (2003).

related infrastructure. In this front, India is engaged in nuclear technology transfer with states such as France, USA and Russia. Similarly, at the level of delivery system, today India has the nuclear capable ICBM Agni-5 to its credit. Besides a robust rocket and missile delivery system, INS Arihant was commissioned as India's indigenous nuclear capable submarine in 2012. At the policy level too, as mentioned, India continues to reiterate its stance of NFU, maintaining only a CMD (Credible Minimum Deterrence) for defence as well as continuing moratorium on further nuclear testing. Indeed, with time, India has become more resolute to write its own rules into a nuclear regime that it considered biased, while also finding ways to navigate its way into the international nuclear club. From Nehru, all the way through to Atal Bihari Vajpayee and Manmohan Singh, when India's position as a nuclear power became overt, New Delhi has tread an interesting path in developing its nuclear program. As explained, 'India's nuclear disarmament diplomacy has moved away from moral and ethical considerations to highlighting national security considerations' (Patil and Vishwanathan, 2022).

Why this book?

The monumental work of George Perkovich, *India's Nuclear Bomb: The Impact on Global Proliferation* (1999) is one of the best works that gives the comprehensive story of how India became openly nuclear in 1998. But the timeline stops there. Raj Chengappa's *Weapons of Peace: The Secret Story of India's Quest to be a Nuclear Power* (2000) falls into the same genre of nuclear history and strategic studies. There are many others in India who have written exclusively on issues of nuclear matters, mostly from a security point of view. Experts such as Achin Vanaik, Brahma Chellaney, Itty Abraham, Bharat Karnad, Uday Bhaskar, Robert Anderson, C Raja Mohan, among many others, come to mind when we think of India's nuclear program and nuclear policy related issues. Amitabh Mattoo and David Cortright's work, *India and the Bomb: Public Opinion and Nuclear Options* (1999), is another meaningful addition that provides a pool of data on how the nuclear tests were perceived by people in India. The surveys, aimed at gauging public opinion (Cortright presents his data in a separate paper written in 1996 that gauges the elite opinion), were rendered useful for policy making on nuclear issues later. On the other hand, scientists have regularly guided policy pertaining to India's nuclear technology and programme.

Having said that, the literature on the subject remains inadequate in its attempt at analysing the role of people – whether scientists, statesmen, diplomats or strategists – who have relentlessly contributed in catapulting India into the nuclear map of the world. When Homi Bhabha is celebrated as the father of India's nuclear program, it is a tribute well deserved, but one that is limited in its understanding. The nuclear program is the making of many such fathers and sons, sisters and daughters of India who find sparse references in the field of nuclear study in India. Similarly, when the United Progressive Alliance (UPA-I) garners credit for furthering India's positon in the use of nuclear power, it only provides a macro image of the progress India has made in the nuclear world. How that progress has been made, is scarcely documented. Who are the people that worked (and are still working) towards making this progress a reality, is even poorly so.

The point is, how many of us are aware of the life struggles and experiences of people who have made India's achievements in the field of nuclear science possible? Of course, we know how Homi Bhabha and Vikram Sarabhai were among the pioneers, encouraged by statesmen like Nehru at the time. We learn about them through scattered sources and their biographies. We know how Indira Gandhi approved nuclear tests in 1974. We may stretch a bit more and include the names of Abdul Kalam or AB Vajpayee for instance, to the list of people whose names are familiar in the context of India's nuclear policy and programme. They are among the more popular personalities. Yet, we lack a body of work that is focused on people as the ontological core of stories related to India's nuclear history. Robert Anderson writes in his book *Nucleus and the Nation: Scientists, International Networks and Power in India* that the definition of who comprises nuclear '"community" is stretched widely here, beyond physics, to include most of those people who considered themselves part of the scientific community. Thus, not just mathematicians and astronomers but also doctors, engineers, lab technicians, industrial and medical researchers, and technologists were included by definition in India's scientific community' (2010, p. 5).

More recently, under a joint project by the Atomic Heritage Foundation and Los Alamos Historical Society, a series of interviews have appeared, titled 'Voices of the Manhattan Project'. It is a collection of oral histories, and shares the experiences of the people associated with the Manhattan Project that gave birth to US's nuclear programme and the race for nuclear weapons in the 1940s. Similarly, in India, we have some biographies (of Homi Bhabha, AB Vajpayee and many others) as well as autobiographies (for example, Raja Ramanna's, *Years of Pilgrimage*, 1991) that give us a peek into how people have achieved what they have in context of India's nuclear story. But we are left wanting to know more, to get a more complete picture.

In early 2022, a show titled 'Rocket Boys' was streamed on Sony Liv and was followed by a second season in March 2023. Directed by Abhay Pannu, it managed to captivate the interests of a large audience, mostly the youth. It is a great attempt to deal with a topic, through a medium that resonates with the millennials and the Gen Zs, as a reviewer of the show put it. But there is a need to compliment these stories through the medium of books as well. As it is, the series faces the same problem that the book attempts to address, i.e. it eulogises the select few.

There is a need for reading the story of nuclear India under a new light, a light that helps us glimpse into the lives of the people who have toiled in its making and how they managed to do so. If scientists and physicists are the creators of nuclear devices and technology, and statesmen the executioners of the nuclear program; strategists are the planners and diplomats the negotiators in the world of nuclear politics. Further, there are the others who labour at various levels to provide manual support to the higher echelons in the field, the stories of whom remain unsung and efforts disregarded. In fact, even the lives and struggles of the popular personalities remain confined to narrations at annual celebratory events or in the works that are singularly biographical.

The nuclear history of the world is markedly different from that of India. It is not only an interesting journey to trace the making of nuclear India but also an endeavour open to debate since no law or act guides nuclear policies in India. Determining what and who shaped nuclear India is an exercise that enriches one's experience as a scholar and reader.

This volume aims to read the developments of India in the field of nuclear science and strategy through the lives of some of the personalities. By narrating their life experiences at a professional as well as personal level, it brings to light the paths taken by scientists, statesmen, strategists and diplomats, the struggles they faced on this path and their influences and inspirations. In fact, the book finds itself segregated into the aforementioned four sections, without however following a chronological time frame. Most importantly, it is an effort to put together the stories, anecdotes, failures and achievements of the people who shaped India as a nuclear power. And indeed, it will be a lesson one has to continue to follow as history progresses.

REFERENCES

- Amitabh Mattoo and David Cortright, *India and the Bomb: Public Opinion and Nuclear Options,* University of Notre Dame Press, 1996.
- C Raja Mohan, *Crossing the Rubicon: The Shaping of India's New Foreign Policy,* Viking, 2003.
- George Perkovich, *India's Nuclear Bomb: The Impact on Global Proliferation,* University of California Press, 1999.
- Indian Academy of Science, URL: https://www.ias.ac.in.
- Raj Chengappa, *Weapons of Peace: The Secret Story of India's Quest to be a Nuclear Power,* Harper Collins, 2000.
- Robert Anderson, *Nucleus and the Nation: Scientists, International Networks, and Power in India*, The University of Chicago Press, 2010.
- Raja Ramanna, *Years of Pilgrimage: An Autobiography*, Viking, 1991.
- Sameer Patil and Arun Vishwanathan, 'India and Nuclear Disarmament: A Quest Rooted in National Security Considerations', *India Quarterly*, Vol. 78(2):261-279, 2022.
- *Selected works of Jawaharlal Nehru,* Series 1 Vol. 8, URL: https://nehruselectedworks.com/pdfviewer.php?style=UI_Zine_Material.xml&subfolder=&doc=January_1937-June_1938-Series1-Vol8.pdf|12|932#page=1.

SECTION A

SCIENTISTS

HOMI BHABHA
The Father of Nuclear India

Rajarshree Narayan Chowdhury

HOMI JEHANGIR BHABHA, a name of prominence on the nuclear stage, is widely regarded as one of the greatest physicists produced by India till date and is accordingly known around the world. His work with Ralph Fowler and Paul Dirac, and association with the leading scientists of the day such as Enrico Fermi, Wolfgang Pauli among others is certainly enviable to any scientist or academician. Bhabha was able to break many stereotypes – not just through his academic career, but also in his various capacities as an administrator, organizer and a first rate leader.

This chapter brings out Bhabha's early attempts to frame the atomic energy policy of India and the opposition he faced in this regard. Also discussed are the negotiations Bhabha entered into with various states to 'jumpstart' the Indian Atomic Energy program. It would be in Bhabha's watch that agreements would be signed with Canada, England and the United States.

Reactors at Tarapur and Trombay were built during his tenure as President of the Atomic Energy Commission (AEC).

Bhabha's role in the international scene is not something that can be glossed over – he was appointed the President of the Scientific Advisory Committee (SAC) at the first ever Geneva Conference on the Peaceful Uses of Atomic Energy, a position he used to put forward the views of the developing states. Bhabha's work in the founding of the International Atomic Energy Agency (IAEA) is also mentioned here. In this chapter, an assessment of Bhabha's work is presented along with his views during the latter part of his life – where he advocated for the peaceful uses of the Atomic Energy but did not completely disavow the potential military applications of the atom – especially following the Chinese nuclear tests of 1964. Although Bhabha's untimely death in 1966 did leave a void in the nuclear department, he had done his job by pushing Indian interests on the world stage.

Childhood and early education

Homi Bhabha was born on 30 October 1909 to Hormusji Bhabha and Meherbai. His father was a barrister and a legal advisor to the Tatas. He also served in the Board of Directors of many Tata companies. Hormusji Bhabha himself was educated in Oxford and also served in the judicial service in the state of Mysore. Bhabha was educated at the Cathedral and John Connon High School in Bombay. As the biographer G Venkataraman puts it:

> As a boy Homi was attracted to playing with Meccano as it

> challenged his abilities. He was also a voracious reader and his reading was helped by his father's wonderful collection of books on various subjects, particularly art and music.
>
> (Venkataraman G, *Bhabha and His Magnificent Obsessions*, 1994, p.17)

From his early days itself, Bhabha had many opportunities to meet national leaders, including the likes of Mahatma Gandhi, among a host of others and learn from them. By the age of fifteen, Bhabha passed the Senior Cambridge Examination. He displayed a strong liking to the subject of science, and as some speculated, by then, Bhabha had already read the theory of relativity! He then entered the Elphinstone College and later the Royal Institute of Science in Bombay before beginning his journey in Cambridge. His admission into Cambridge would also bring him into contact with the most formidable minds in the field of Physics and set him off on his destiny.

The European Connection and Entry into Mainstream Discourse

Bhabha the 'colonial scientist' or the 'western scientist'?

Itty Abraham, in his work, *The Making of the Indian Atomic Bomb: Science, Secrecy and the Postcolonial State,* talks about the difference between what George Basalla calls a 'Western Scientist' and a 'Colonial Scientist'. In case of the former, there is what one may refer to as 'originality' in terms of the theoretical formulations and breakthroughs; while the latter may be seen as no more than 'derivative practitioners' who are considered no more useful than those who gather data, but are considered intellectually incapable of breakthroughs in the theoretical field:

> [I]mpression of colonial scientists as derivative practitioners, not quite up to the standards of their metropolitan colleagues, valued more for the data they provide than for the theoretical breakthroughs ...once scientists enter the frame, 'colonial' becomes euphemism for inferior.
>
> (Abraham, 1998: 35)

Homi Bhabha began challenging this hierarchy during the 1930's while in Europe. It was not just the academic career, or the charming and suave nature that drew admiration from men and women alike, but the embodiment of 'Western Culture' that was manifest in his ways. Abraham writes:

> Everyone in Europe was curious about Bhabha. Unlike his Indian contemporary astrophysicist S Chandrasekhar, who was self-effacing to the point of invisibility, Bhabha was not shy about engagement with Europeans...Bhabha would strike up conversations with strangers in trains about the books they were reading.
>
> (Ibid: 38)

The physicist Hendrik Casimir (whose wife was among the many women who were charmed with Bhabha – in fact referring to him as a 'fairy tale prince') likened him to the savage in Aldous Huxley's *Brave New World* (1946), where the savage learns English by reading William Shakespeare but speaks archaic that is both out of place and out of time. Similarly, for Casimir, Bhabha was a physicist but his roots somehow seemed to betray him and were 'unscientific'.

> What Bhabha cast into doubt was the very difference between colony and metropolis that is built into the institutions of science...Casimir could admire Bhabha's given qualities...as a way of reminding us that Bhabha's mind was 'unruly' as befitted the idiot savant he really was...
>
> (Ibid: 41)

To look at it another way, in the words of Frantz Fannon:

> The White man's eyes break up the black man's body and in that act of epistemic violence its own frame of reference is transgressed, its field of vision disturbed.
>
> (Bhabha. HK, 'Interrogating Identity', 1994, p.185)

Therefore, for the Western Scientists, there was a sense of 'deep ambivalence' rooted in knowing that Bhabha was a 'colonial scientist', but at the same time, their physical encounters with a man, who in no ways was less in any sphere, struck a serious blow at the heart of hierarchic ordering of scientists:

> [T]he difference of Western Science lies outside the field of science...it depends on an act of power to know that it is dominant.
>
> (Abraham, 1998:41)

Bhabha's academic career in Europe

Homi Bhabha left for Europe in 1927, having reached the age

of eighteen. Once there, in England, he enrolled in the Gonville and Caius College in Cambridge. This was not unexpected as his uncle Sir Dorabji Tata had been a student of the same college as well. In fact, the mathematical acumen shown by Bhabha in his early years had stirred high hopes in the minds of both his Father Sir JH Bhabha as well as uncle Sir Dorabji Tata that he indeed was the ideal candidate for heading the Jamshedpur Steel Industries of the Tatas in the future. Accordingly, therefore, he had enrolled for the Mechanical Sciences Tripos.

However, Bhabha joined Cambridge during what may be defined as revolutionary times – a time that may be best described as the Golden Era of Physics i.e. developments were in full flow in theoretical Physics under the likes of Heisenberg, Schrodinger, Dirac to name a few. Equally proficient were the likes of experimental physicists with the contributions made by Lord Rutherford and the Cavendish Laboratory. Homi Bhabha, having attended the lectures of Paul AM Dirac, realised that his heart lay with Physics and not in Engineering. In a letter, dated 8 August 1928, addressed to his father, Bhabha wrote:

> I seriously say to you that business or a job as an engineer is not the thing for me. It is totally foreign to my nature and radically opposed to my temperament and opinions. Physics is my line...I am burning with a desire to do Physics...I have no desire to be a 'successful' man or the head of a big firm. There are other intelligent people who like that and let them do it...I therefore, earnestly implore you to let me do physics.
>
> (Venkataraman, 1994:18)

Initially reluctant to let the son have it his way, the father, however, struck a deal where he would let Homi pursue Mathematical Tripos if and only if he were able to secure a First Class in his Engineering Tripos. Homi Bhabha did not disappoint and did exactly what his father asked, securing a first class in Mechanical engineering in 1930 and again during his Mathematical Tripos in 1932.

With his achievements in the Mathematical Tripos, Bhabha earned the Rouse Ball Travelling Studentship which allowed him to travel to Copenhagen and Zurich where he could have discussions with people like Niels Bohr and Wolfgang Pauli. In fact, Bhabha presented a letter of reference to the latter from Ralph Fowler back in Cambridge. In the letter, Fowler had equated Bhabha to a spoiled child and likened Pauli's institute in Zurich to a boarding school. The tone as well as the content of the letter were highly amusing to Wolfgang, such that he took an almost instant liking to Bhabha.

Back in England and Cambridge, around 1936, Bhabha worked with the Senior Physicist Walter Heitler on the issue of cosmic ray showers and developed what is now the Bhabha-Heitler Cascade Theory. Bhabha continued to apply Dirac's equation to electrodynamics, and according to Lord Penney, one of Bhabha's biographers, in a way, Bhabha's experimental work was important in validating the success of Dirac's theory.

Around 1938-39, Homi Bhabha applied to the position of Reader in Liverpool University (although some believe, among them the biographer Chintamani Deshmukh, that he was offered the position by James Chadwick). He was however turned down. The reasons given ostensibly pointed to Bhabha being over qualified for the job, but as Itty Abraham was to point out later:

> The figure of the colonial scientist was no less confusing for the young students of Liverpool. How could they be taught physics by a colonial subject? Simply put, if Bhabha was good enough to teach them Physics, couldn't the Indians be good enough to run their own country?
>
> (Abraham, 1998:42)

Having been turned down at Liverpool, Bhabha decided to come back to India for a vacation in 1939, and it was then that the Second World War broke out.

Bhabha settles in India

Prior to leaving Cambridge, Bhabha had secured a Royal Society grant to work in experimental physics in the Laboratory of Patrick Blackett. However, with the outbreak of the war, there was a complete stop to the entire operation. In fact, the situation was such that the once fluid exchange across borders gave way to something far more sinister in the form of devotion of energies to the war effort. Secrecy was of paramount importance as 'continental physics' was abandoned in favour of 'nationalized physics' (Ibid: 42).

While Bhabha was looking forward to contributing towards the Allied war cause, he was repeatedly turned down owing to his 'colonial national origins'. His being part of a population that was under rule by a foreign power, at least covertly, meant that he could never be seen as anything beyond a 'colonial scientist'. This, despite his achievements in the field of science in Cambridge, continued to haunt him, though in no way did it break his resolve to keep contributing to science.

In the first year of his stay, he divided his time between Bombay and Bangalore. At Bombay, Bhabha delivered a series of three lectures on Cosmic Radiation, while two of his research papers were published in the *Proceedings of the Indian Academy of Sciences*, Bangalore. Bhabha had also established a certain degree of rapport with CV Raman, who encouraged the former to join the Indian Institute of Science (IISc). Raman had grand plans for the Department of Physics; he was particularly keen on nuclear physics and cosmic rays. Bhabha was just the man for the job.

Subsequent works of Bhabha through 1940 had the address of the physics department at the Indian Institute of Science, Bangalore. Bhabha's first joint paper in India was with a Mathematician BS Madhava Rao on the 'Scattering of Charged Mesons', published in a journal by Indian Academy of Sciences in December 1940. The year 1942 proved especially remarkable for Homi Bhabha as he was awarded the Adams Prize from the University of Cambridge for his thesis, 'The Theory of Elementary Particles and their Interactions'. Further, Bhabha was appointed as a fellow of the Royal Society, London, in 1941. He was also, at this time, promoted to full professorship in Cosmic Ray Research at the IISc. Unfortunately, however, this was also the time when Bhabha's father passed away, taking a toll on his personal as well as professional life.

On another note, in 1944, with help from the 84th Air Depot of the US Air Force stationed at Bangalore, Bhabha tested airplanes to fly at five thousand to thirty thousand feet altitude to measure the high altitude mesons. The results of this were published in two papers appearing in 1945 and 1946. The specialty of doing science in India, thus, was pointed out by Itty Abraham:

> The obvious attraction of the manipulation of nature to provide a solution to major scientific problems took on a distinct nationalist hue in India. It was assumed that Indian Scientists could not compete with their Western counterparts for funds to build large experimental facilities. Hence, conducting big science experiments through this 'reading' of the Indian landscape was represented as an especially creative nationalist response to the lack of resources compared to the West.
>
> (Ibid: 45)

It is clear that Bhabha, while gaining most of his formative education in Cambridge to the extent of blurring the line between what Itty Abraham refers to as the colonial scientist versus the Western scientist, was now to turn towards the advancement of science or more specifically nuclear science in India.

Bhabha – the Institutional Builder of Nuclear India

Bhabha was dissatisfied with the scientific community in India – the fact that they did not present a united face to the world.

For example, the friction between CV Raman and Meghnad Saha was stark by then, but was not particularly endorsed by Homi Bhabha. In other words, between the Indian Academy of Sciences (IAS) of Raman, and National Academy of Sciences, India (NASI), of Saha, it was becoming difficult to project a unified face of Indian Science; added to this was the politics that further depressed Bhabha, as his biographer Deshmukh recounts:

> About his rejection of the Allahabad offer, he wrote to Birbal Sahni that he was not interested in university politics or in bossing over others, but he would like no interference from others too...he had a long meeting with the Vice Chancellor of Allahabad University...but he did not find the atmosphere in the physics department to his liking and so rejected the offer.
>
> (Deshmukh, 2010:18)

Yet, Bhabha was now beginning to appreciate the potential that science and technology held for India, beyond the cynicism of professional or personal aggrandizement, especially with reference to atomic energy. But first what was needed was financial support. Bhabha approached Sir Dorabji Tata Trust for it. He had gained enough knowledge regarding the functioning of the scientific institutions to be confident that an institution of such sort would have a bright future in India.

Establishing the Tata Institute of Fundamental Research (TIFR)

Before turning to his family contacts, Bhabha had looked forward to support from the CSIR (Council of Scientific and Industrial Research), headed by the eminent scientist Shanti

Swarup Bhatnagar, with whom Bhabha shared a good rapport. Bhabha made his plans and intentions apparent in a letter to the Chairman of the Sir Dorabji Tata Trust, Sorab Saklatvala. In the letter, Bhabha pointed to the lack of any substantial institution dedicated for specialized research, particularly research in nuclear physics, which he believed to be the future of India. He wrote:

> Moreover, when nuclear energy has been successfully applied for power production in, say, a couple of decades from now, India will not have to look abroad for its experts but will find them ready at hand... The subjects on which research and advanced teaching would be done would be theoretical physics, especially on fundamental problems and with special reference to cosmic rays and nuclear physics; and experimental research on cosmic rays.
>
> (Letter from Homi Bhabha to Sir Sorab Saklatvala, dated 12 March 1944)

Further, Bhabha hoped to assuage any doubts that the trustee board might have regarding the control of the proposed institution when he stated:

> [F]inancial support from the Government need not however, entail Government control…Many of these independent scientific institutes in Great Britain nowadays are receiving a substantial state support, but nearly always…a buffer of some kind is interposed to prevent Government support from becoming Government control.
>
> (Ibid.)

On a concluding note, Bhabha also mentioned a 'personal front' where there was always an option present of returning to Europe to continue with research, but that he was convinced about the duty of an individual to stay back in one's own country and 'build up schools comparable with those that other countries are fortunate in possessing'. The Tatas responded promptly and affirmatively while also suggesting that help could be sought from the Government of Bombay Province and University of Bombay, to which Bhabha agreed. With this began Bhabha's efforts in building and harnessing India's scientific temper that eventually attributed to India's successes in the nuclear field and beyond.

The most important challenge then was the recruitment of faculty that would provide a good head start. The first person Bhabha approached was S Chandrasekhar. Though initially Chandrasekhar responded to the offer positively, he chose to stay back in the US and continue teaching at the University of Chicago. Bhabha reached out to other senior academicians of the time that resulted in positive outcomes; one was DD Kosambi, a Harvard graduate with research in mathematics and statistics, who was also known among the academic circles for his knowledge in Sanskrit and for being a Marxist scholar. RP Thatte, who at the time was teaching electronics in SP College in Pune, came on board as well. This gave fresh life to Bhabha's visions; and the institute, christened the Tata Institute of Fundamental Research (TIFR), started to function in an official capacity on 1 June 1945.

The TIFR budget for the fiscal year was set at Rs 80,000, of which Rs 45,000 came from the Dorab Tata Trust; Rs 25,000 from the Government of Bombay; the remainder from the Atmospheric Research Committee of the CSIR. The Institute was formally inaugurated by Sir John Colville, Governor of Bombay Province, on 19 December 1945.

TIFR developed as a premier institution that reflected on India's growing scientific needs as a modern nation of the new emerging world order. Researchers from other Universities soon gravitated towards TIFR, and within four years, it had expanded to an area six times larger than its initial allotted space. In 1955, a Tripartite Agreement was signed between the Dorab Tata Trust, the Government of Bombay and the Government of India that led to administrative and financial changes in the TIFR structure. Support was now to come from the Department of Atomic Energy (DAE) and grants mostly directly from the Centre. This alleviated the importance of TIFR for research in India, so much so that by the time Homi Bhabha died in 1966, TIFR received an allocated budget of Rs 1.5 crores!

Work underway on the Indian Atomic Energy Programme

With the process of India's independence well underway, the CSIR established the Atomic Research Committee which was to be an important step in the setup of the nuclear research programme. On 26 August 1947, the CSIR set up a 'Board of Research on Atomic Energy' with Bhabha as its chairman. He, however, wasn't satisfied with the board and its limitations, thus writing to Prime Minister Jawaharlal Nehru on 26 April

1948 – 'On the Organisation of Atomic Research in India'. The note was to prove crucial on which Nehru acted promptly, 'accepting every major proposal of Bhabha'. It is said:

> At the outset, Bhabha made it clear that the quickest and the most desirable way of developing Atomic energy in India would be to come to an agreement with the Government or Atomic Energy agencies of one or more countries; India needed to build an atomic pile to quick-start its own self-reliant atomic energy programme based on the plentifully available thorium...He then put down two basic conditions that virtually dictated the organisational structure for Indian Atomic Energy Research...
>
> (Deshmukh, 2010:31)

The conditions that Bhabha put forward were largely to lay the template for most of the work pursued subsequently in the field. Notwithstanding the proposed efficiency of the structure, it was to be embroiled in controversy as critics would often point towards the cloak and dagger nature of its operations versus the objectives that were to supposedly guide the workings. This led to the conceiving of the Atomic Energy Commission (AEC) with its own secretariat, independent of any other Ministry or Department of the Government.

The commission was to be entirely composed of scientists – the Chairman, the Director of the CSIR as the member-secretary, and one other 'eminent scientist' FRS (Fellow of the Royal Society). Apart from this, Homi Bhabha also asked for a grant of Rs 1 Crore to be spent across the next three to five years which interestingly was requested all at once, as part of a detailed budget! Prime Minister Nehru accepted all the major suggestions of Bhabha. Thus was formed the Atomic Energy Commission (AEC) on 10 August 1948 with Bhabha as the Chairman, Shanti Swarup Bhatnagar as the Member-Secretary, and KS Krishnan (who was the then director of the National Physical Laboratory) appointed as the third member. There were no bureaucratic involvements from any other department or ministry. Bhabha was answerable only to the Prime Minister. Indeed, Bhabha had created a niche for himself much to the envy of others.

The Atomic Energy Commission (AEC) starts working

Such was the stature of both Jawaharlal Nehru as well as Homi Bhabha that the programme from the outset could be questioned by very few. It was understood that the project was time consuming, the gestation period was long, but the wait would be worth it, for the product would most truly be Indian – a grand achievement of a newly independent state catapulting it into the same league as the 'big players' in the world.

Admittedly, no effort was spared for the purpose. A government owned corporation, Indian Rare Earths Limited (IREL) was set up, while a production plant was started at Alwaye, Kerala in 1952, for the processing of monazite sand. Earlier, in 1946, the Travancore government was negotiating a new deal with British Titan Products and DuPont to construct a titanium plant, while at the same time making arrangements to sell monazite sand to other foreign agencies. This was considered a genuine problem because monazite, in layman's terms, was the raw material for thorium and was held in great importance for the most obvious of reasons. Both Bhabha as well as Nehru lobbied hard to prevent what the Travancore government had in mind from materializing. This indeed was one of the earliest indicators of the lengths to which both Bhabha and Nehru were willing to go to maintain pre-emptive control over anything that could be used in the nuclear project. Nonetheless, uranium deposits were discovered at Jaduguda in Bihar, while the work on the thorium plant was almost immediately begun in earnest at the fishing village of Trombay in 1953. The following year proved significant for Bhabha. But suffice it to say that since the inception of the program, the progress was kept under wraps, and information only allowed to percolate on a need-to-know basis. Not surprisingly, it was clear that not many needed to know!

Formation of the Department of Atomic Energy (DAE) and the challenges faced

Bhabha continued to dominate the nuclear program of the country unparalleled. But this did not mean he was not confronted or questioned at times. To his credit, however, he managed to skilfully manoeuvre the situations and remain a step ahead of his critics.

Since its inception, the AEC had expanded so much that it became necessary for a separate department to be formed in the Central Government to complement the rapidly developing programme. Consequently, in August 1954, the Department of Atomic Energy (DAE) was formed. Bhabha was appointed the Secretary, answerable once again solely to the Prime Minister. The Department had all the authority to fund, create, manage and operate all aspects of the Atomic Energy Programme. On the heels of the DAE, was formed the Atomic Energy Establishment, Trombay (AEET), whose Director, unsurprisingly, was none other than Homi Bhabha.

Despite all these achievements of Bhabha, he did face challenges, most evidently in the form of criticism from Meghnad Saha, who had impeccable records, both as a scientist as well as a nationalist. Like Nehru, Saha tended more towards the socialist ideology, and both were what may be referred to as unmitigated votaries of large-scale industrialization. They were in close concert initially, but drifted apart as Nehru grew close to Bhabha. Saha was being increasingly marginalized from the nuclear program of the country. Earlier, he had challenged Shanti Bhatnagar over the construction of oil refineries, especially given that they were from the UK among many other states. As the eminent academic Robert Anderson points out, Saha questioned the validity of import of steel, manufacture of energy, ceiling on electricity rates, and even the government's inattention to archaeology. Yet, conspicuous was the absence of the atomic question in any of his criticisms. On that note, it is worthwhile going into the debate that accrued over the subject, as it was one of the major attacks mounted on the credibility of the AEC and DAE.

As already stated, Saha's position in the nuclear circles was waning very fast. Then, he did something that came as a bolt out of the blue – he contested for the Lok Sabha election from the Calcutta Northwest constituency as an independent candidate, with support from the left parties. To the surprise of all (especially the scientific community) he won! Saha was now determined to do what he could not as a scientist, i.e. launch attacks on the AEC. To quote Robert Anderson:

> He made an indictment of Bhabha, Bhatnagar, and Krishnan, all, like him, Fellows of the Royal Society. He pointed out that there was really nothing unique about atomic scientists, that they were simply ordinary physicists and chemists who had studied the nuclear subject thoroughly, that's all...the atomic scientist is not a new race that has come into existence.
>
> (Anderson, 2010:237)

As if the attack on the nuclear scientists were not enough, Saha sought to undermine the work of the AEC:

> India has got an Atomic Energy Commission, and five years ago it had announced that we were going to set up a nuclear reactor within five years. This is 1954 and nuclear reactor has not been set up...I do not know how long it will take us to make good this proposal. It may take us years to set up nuclear reactor.
>
> (Abraham,1998:73)

Saha, to borrow from Robert Anderson's vocabulary;

> ...crystallized his negative judgment of the atomic energy program at two special meetings...The first of the two special meetings, held in September 1954 at Saha's request, was at Nehru's house.
>
> (Anderson, 2010:237)

At the meeting, Anderson notes that Saha led the criticism well, in presence of personalities like Nehru and Bhatnagar, pointing to the extreme secrecy as being both unnecessary as well as unfair to the scientific community in general. Saha, then went on to suggest that India should seek to emulate the French model that sought complete self-reliance, but disavowed secrecy. Suggestions were also made that scientists and politicians should be roped in to give shape to the nuclear policy of the state, while the administrators of the Atomic Energy Programme were to be full-time employees who were to have no other jobs. It must be said that Homi Bhabha was not present at this meeting, but a second round was scheduled for 26 and 27 December.

The meeting was held at the National Physical Laboratory (NPL) and referred to as the Development of Atomic Energy for Peaceful Purposes in India, and the objectives, in the words of Robert Anderson were;

> [T]he preparation of a National Policy Statement for the forthcoming Geneva conference on peaceful uses of Atomic Energy and to help sort out political differences in time for the AEC to make choices in pending reactor negotiations.
>
> (Anderson, 2010:238)

Itty Abraham (1998) assesses the situation by pointing the following:

- Choice of the National Physical Laboratory as the site for the meeting could hardly be called unbiased, as the then director of the NPL was KS Krishnan, a protégé of CV Raman and not what one may call an 'impartial adjudicator.'

- There was every attempt to buffer the AEC from criticism. The meeting became a platform to share information rather than critically assess the appropriateness of the AEC's ways in the development of Atomic Energy.
- Besides, the papers presented were 'vague generalities about the value and importance of atomic energy for development'; the scientists were mainly drawn from the TIFR and AEET.

As a consequence, the day could not be saved for Saha, since there was very little impact made to the overall workings of the AEC or the DAE. The fact remained that despite all his scientific and national credibility, Saha was unable to do much.

Reactors and help from abroad

Bhabha had won a major victory against his nemesis Meghnad Saha, but all was not well. Deep down, Bhabha was aware that tangible results needed to be produced if other critics were to be kept at bay. In other words, a certain urgency had crept into the workings of the AEC. A reactor needed to be built. There was also the matter of pride to be put aside by the scientists in order to build that reactor. They would now have to become more pragmatic and seek help from abroad. No matter how much it ate at them, there was ostensibly no other option.

For Bhabha, association with the Western states was no more than a means to an end. Bhabha sought to use the natural reserves of the country such as thorium, beryllium and manganese. They were to be offered to select foreign countries in exchange for their cooperation, thus furthering the atomic energy programme of India. In this endeavour, Bhabha counted on the support of Bhatnagar, who ensured that no interference came from the scientists at CSIR (technically, the AEC was still a part of the Department of Atomic Energy at CSIR). Also, in 1952, Bhabha had written to Gordon Dean, in the latter's capacity as the Chairman of the US Atomic Energy Commission, asking for all de-classified information on reactor theory, design and technology. He was even very specific and referred to the Harwell Reactor in the UK and the Heavy Water Reactor at Chalk River in Canada. With this, Bhabha's visions of operationalizing nuclear processing in India began to take shape.

The British Connection and APSARA

As early as 1948, Bhabha had begun to approach his old colleagues in the UK. The latter, however, were not willing to test the limits of US disapproval, given the sensitive geopolitical context following WWII, which meant that they turned down Bhabha's request. However, Bhabha responded:

> [I]f (cooperation is not possible) then we shall clearly have to follow other roads which we know are open to us, leading to the same objectives.
>
> (Abraham, 1998:83)

The boldly worded letter, which almost read like an ultimatum, jerked the British out of whatever lull they were in. Thereafter, the likes of John Cockcroft, Henry Tizzard and Edwin Plowden were all in favour of helping the Indian

Atomic Energy programme. Bhabha as well as Bhatnagar had a series of meetings with the British officials through 1954, where requests were made for helping with the building of a reactor and converting uranium ores into metal for fabrication. Bhabha also made the request for export of five tons of heavy water. The British obviously did not relent, citing their own domestic needs. However, Cockcroft came up with an idea that was rather ingenuous. He suggested that Bhabha build what may be referred to as the 'swimming pool' type reactor, whose capacity would be 1 MW. It was to be situated in Trombay.

> It was to be a light-water enriched uranium reactor, on the lines of one at Harwell, UK…An agreement was signed in October 1955 between the UKAEC and the DAE. Except the fuel, the rest of the reactor was to be designed and built totally indigenously... The assembling was started in 1955 and within a year was commissioned. It attained criticality on 4 August, 1956. It was the first nuclear reactor in Asia.
>
> (Deshmukh, 2010:35)

Thus, the reactor named APSARA, attaining criticality within a year, killed two birds with one stone. Not only did it serve to silence the critics of Bhabha and the Atomic Energy Programme, but also, notwithstanding the British help at crucial aspects, raised confidence for the scientists who were involved in the project.

The Canadian-Indian Reactor Project

The Scientific Advisory Committee was formed under the auspices of the United Nations, following the Atoms for Peace Programme by the US President Dwight D Eisenhower in 1953. The SAC was composed of scientists drawn from seven states, namely India, Brazil, Canada, USA, USSR, France and UK. The conference was held in Geneva in the first week of July 1955, where Bhabha was appointed the President, a decision that caused waves of jubilation across the country. Although the conference did not start off on the highest of notes, it soon became clear that no one country had gained ascendency in definitive terms over the others. Neither had any country made enough breakthroughs to warrant the unabashed use of nuclear energy for civilian purposes.

What stuck out, however, was the Canadian offer to the Indians in the form of an NRX-type 40 MW reactor. The reason behind the offer was of course Canada's willingness to carve out an independent space for itself in the emerging world order and not merely be seen as a lackey to the US and its interests. Therefore, from a purely pragmatic point of view, Canada was also looking forward to expanding the market for their product. India proved to be the ideal client.

Collaborating with India at this level could have been the idea of Nik Cavell or William J Bennett – the former was the Canadian administrator of the Colombo Plan while the latter was the head of the Atomic Energy of Canada Limited (AECL). On 5 April 1955, the Vice President of the AECL, WB Lewis, wrote to Bhabha expressing the Canadian interest in helping the Indian Atomic Energy Programme. But Bhabha proceeded with caution and let the Canadians know that he preferred the British reactor while also considering the US-built swimming pool reactor. Thereafter, the Canadians sweetened the deal and agreed to supply extra funds, in fact, going to the extent of agreeing to foot half the expenses of building the reactor. The heavy water needed for the project was to be supplied by the US AEC.

The construction for the CIRUS (then known as CIR i.e. Canada-India Reactor) started in 1956.

> In August 1956, the north site of Trombay was a bush jungle...and by this time a tractor was beginning to clear the site proposed for CIR. Meanwhile, recruitment of engineers was in full swing and within a couple of months, a group of about 40 young engineers were assembled under V Surya Rao who earlier had experience working in power station[s] in England. Some of these recruits were sent to IISc, for a short course on power engineering.
>
> (Venkataraman, 1994:176)

For the CIR, Homi Bhabha wanted as many parts as possible to be made in India. Unlike the APSARA, he felt the CIR ought to have Indian made fuel elements. The uranium rods, similar to the ones used in the NRX, were to be made in India. While the Canadians were initially sceptical about the plans, Bhabha offered them the use of the fuel elements for about a year. Only when they performed as well as Bhabha said they would, were the doubts of the Canadians laid to rest.

For Bhabha, CIRUS was necessary for the following reasons:

- As a training ground for reactor engineers, especially operators. Later, they would operate our nuclear power stations.
- As a powerful source for producing various isotopes needed in medicine and industry.
- As a source of plutonium. The irradiated uranium rods would contain quite a bit of plutonium. By suitable chemical processing, this could be extracted.
- As a powerful neutron source for various experiments, particularly in physics.

The CIRUS project was finally up and running by the year 1960, with operational testing taking a few months. In fact, the Uranium Metal Plant was operational by mid-1959, thereby meaning that half the initial charge used in the CIRUS came from Trombay's own production. Finally, the reactor attained criticality on 10 July 1960. With the APSARA, and now the CIRUS, having attained criticality, Bhabha set his sights on the creation of a fully indigenous reactor – the ZERLINA. As Deshmukh puts it, 'a zero-energy research reactor for lattice investigations and study of different fuel assemblies was undertaken'. ZERLINA was made a reality soon enough and attained criticality on 14 January 1961.

Negotiations with the United States

To begin with, the United States appeared aloof and high-handed in its attitude towards sharing nuclear technology with any of the other states. It was their belief that they were the progenitors of the technology and thus had almost exclusive rights to it. However, as Itty Abraham states:

> By the end of the 1950's, reflecting the altered international context, and in some contrast to their attitude a decade before, the US atomic energy establishment was now keen to get involved in India.
>
> (Abraham, 1998:93)

As the case was, until 1959, the US was only involved in the capacity of fuel supplier in the form of the twenty-one tons of heavy water they had supplied for the CIRUS project. Bhabha now approached the US with the proposal to build a 250 MW capacity plant in India. The United States cited the capital costs as a major hindrance towards the development of nuclear plants (cost per kilowatt for the former was estimated at about $335, against an estimated cost of about $180 per kilowatt for energy production from conventional sources). Bhabha, however, had been expressing concerns during meetings with the US agencies throughout 1959, where he spoke of India's development programme and the role that atomic energy was to play in it. He also underlined the role that the United States may play in furthering the atomic research in India. Further, his second concern revolved around the funding for the collaboration which had to be ideally drawn from other sources, so as to not disturb the alternate source of developmental aid for India. By then, Bhabha could speak with even greater authority than before because of a resolution passed in the Lok Sabha the previous year where the following decisions were made:

- The AEC was to consist both of full and part-time members, totalling to no less than three and not exceeding seven members.
- The Secretary of the DAE was to be the ex-officio chairman of the AEC.
- Another full-time member of the AEC was to be Member for Finance and Administration, who would also hold the designation of the ex-officio Secretary to the Government of India (GOI) in DAE in financial matters.
- The Director of AEET (Atomic Energy Establishment, Trombay) was to be the third ex-officio full-time member in charge of research and development.

Also, the AEC, at the same time, was given the powers to formulate the policy of the DAE and responsibility for getting government approval for the budget to run the latter. Bhabha, thus, significantly consolidated his position, where following on the heels of this, the Atomic Energy Act of 1962 was passed. It included extra powers and responsibilities, especially with reference to radiation protection, power generation and distribution.

However, coming back to the negotiations between the United States and Bhabha, we see that with regards to the global tender for applications to build the first large power reactor of India, the Canadians were leading the race; so was their heavy-water natural uranium reactor that fit the Indian profile closely. However, at the last moment,

> [R]epresentatives of General Electric Company and Westinghouse met (AEC scientists) and explained to (them) that their reactor systems using enriched uranium were much more economic to build and operate...An agreement was signed with the US to build twin 200MW enriched uranium reactors at Tarapur near Bombay…A few months later India received an US $80 million loan from the United States for the construction of the Tarapur reactors.
>
> (Abraham, 1998:94).

An agreement was finally signed in 1963 with the US – with the General Electric (GE) company to be precise.

As we have noted throughout this section, the zeal surrounding Bhabha as an institution builder of nuclear India is not overhyped. His many contributions to initiating ideas, finding adequate foreign support, as well as for being a pillar of strength for the nuclear fraternity in India is well understood. Yet, Bhabha was an organizer, leader, and dare I say a philosopher who lived at a time when India was at its early stage of nation building. Bhabha's role in the development of the International Atomic Energy Agency, his role in the Geneva conference are a must to study if one wishes to gain an understanding of Bhabha as a person. While touching upon these aspects, the next section will bring to light the manner in which Bhabha was able to negotiate at an equal footing with the so called 'big players' in the world atomic stage. The story of Bhabha has always more to reveal than one expects.

Bhabha – The Philosopher, Organizer and Leader

The First Geneva Conference on Peaceful Uses of Atomic Energy (1955)

The stature of Homi Bhabha, both as a scientist as well as a diplomat, would be incomplete, if his involvement and views at the First Geneva Conference is not presented. As one is aware, the Americans were the first to develop nuclear energy, for military as well as civil purposes – a fact that made them fiercely protective of the technology. However, as it became apparent that the British, French, Soviets and Canadians were on their way towards developing indigenous Atomic Energy Programmes, the Americans came up with something ingenious to counter the progress – the Atoms for Peace Programme, which was announced by the American President Dwight D Eisenhower. The objective was to share atomic technology for peaceful purposes among the states to make sure that the power of the atom was not being used for diabolical purposes. To this end, the then American Secretary of State proposed a resolution that was accepted by the UN General Assembly on 4 December 1954. An international atomic agency was established and a Scientific Advisory Committee was formed, where, as already mentioned, seven member states were included, India being one among them. Naturally, Homi Bhabha was chosen as the Indian Representative and became the eventual President of the committee.

The Conference was held between 8-20 August 1954 at Geneva, Switzerland. About 1400 delegates from 73 states attended the conference with various observers and media following it attentively. In total, about 1000 scientific and technical papers were submitted. In the midst of this,

> ...explicitly adopting the persona of the ideal, apolitical scientist, Bhabha was able to situate himself as an observer not invested in the passions of the world, yet aligned within a Western tradition that the audience could not but identify with.
>
> (Abraham, 1998:99)

In his address, Bhabha divided human history into three epochs. The earliest corresponded with the river valley civilizations of the Middle East and South Asia where the characteristic feature was the usage of muscle power, both human as well as animal. The second stage of human development occurred with the expansion of the Industrial Revolution and the expansion in the usage of fossil fuels. And lastly, according to Bhabha, the next stage in human history, related to the 'acquisition by man of the knowledge of how to release and use atomic energy, must be recognized as the third great epoch in human history' (Ibid: 100). In his speech, Bhabha referred to the relation between energy supply and the development of civilisations. The level of development in the first stage was low. In the second stage, the level was better and men were known to enjoy higher standards of living. However, there still were masses who could not become part of the mainstream development discourse. This, coupled with the fact that the known reserves of energy were being exhausted – as per Bhabha's estimate, within the next century (contingent upon the fact that the entire human race consumes energy at the same level per capita as does the USA) – meant that in order to keep the human civilization afloat, it was a must that atomic energy be developed. Bhabha stated that while the scope of the conference did not extend to include nuclear weapons, it was not possible to draw a concrete line between what may be referred to as 'peaceful application' of nuclear energy and diversion of the science for 'weapons purposes'.

In conclusion, Bhabha pointed to the progress which was made in apparent isolation by the various states and expressed hope that the scientific and more importantly the human community could find ways to collaborate in this endeavour for 'common good'.

The International Atomic Energy Agency (IAEA)

As it turned out, the aforementioned Geneva conference on Peaceful Uses of Atomic Energy became the perfect breeding ground to establish the IAEA under the auspices of the United Nations, especially given the fact that the conference talked about spreading the benefits of atomic energy across the world. Some sort of a supervisory body was needed to oversee the application of the technology. To this end, a Preparatory Conference was held in Washington in 1956, involving twelve states, to agree upon the statutes of the IAEA. Later that year, a Plenary Conference was held in New York to discuss the issues regarding controls and safeguards. Another issue which crept up was related to the location of the IAEA. After some discussions, it was finally decided that the location would be

Vienna, something to which the USA and USSR seemingly agreed.

Bhabha was one of the members in the Board of Governors of IAEA. In this capacity, Bhabha was always a strong supporter of the developing states' interests, while his stance on safeguards led him to assert that they should apply equally on states – be they newcomers to the atomic scene or established forces. His perseverance paid off and safeguards were applied to even some of the major power stations in the USA, UK and others. Bhabha's argument also focused on keeping of the safeguards to a minimum where he advocated that major safeguards should be applied to the chemical plutonium separation plants, i.e. where the plutonium is separated from used reactor fuel and can actually be diverted towards the making of a bomb. Bhabha's inputs were indeed crucial.

Nuclear Disarmament

As a founding member of the IAEA, Bhabha had a say in the issue of nuclear disarmament, where his view differed from that of the United States. The US believed in non-entry of other members into the club, freeze or at least non-expansion by the powers established. On the other hand, though Bhabha supported the freeze and the non-expansion of the established powers, he also viewed the existing system as no less than discriminatory. For Bhabha, the use of atomic power was to be kept within the bounds of constructive, i.e. peaceful, but developmental use was to be allowed. As Deshmukh puts it,

> even if widespread use of atomic energy for peaceful purposes

> forced us with political and military problems, we had no option but to solve those problems...A 300 MW power station would produce enough plutonium for production of between 20 and 35 atomic bombs a year, depending on their size.
>
> (Deshmukh, 2010:46)

Bhabha also suggested the creation of a UN Security force which would be strong enough to maintain peace, in what according to him was a relatively disarmed world. Therefore, Bhabha's views on nuclear disarmament defined the future of nuclear use by countries other than the powerful nuclear group.

Conclusion and assessment of Bhabha's contributions

Homi Bhabha's untimely death on 25 January 1966 was a major blow, not merely because the world community had lost a scientist par excellence, but also because India as a state had lost a visionary, a scientist, a leader and an organizer. Although the legacy of Bhabha was carried on under capable hands through the likes of Raja Ramanna, Vikram Sarabhai among others, his absence would always continue to plague India, especially when it came to negotiating on one's own terms in the international stage.

The above is, however, not to suggest that all of Bhabha's plans and ambitions were realistic as well as feasible. George Perkovich points out:

> Bhabha grossly overestimated the economic merits of nuclear power...second major problem was Bhabha's claim that India could build nuclear plants more cheaply than the

> technologically advanced states with their higher labor costs.
>
> (Perkovich, 1998:31-32)

In other words, while the projections of Bhabha were optimistic, not all were right. Further, things began to change following the Chinese tests of 1964. Bhabha pushed his argument for the bomb, while he also floated the idea of what may be referred to as security guarantee from the superpowers. Bhabha lay the onus on the US and USSR to extend formal security guarantees if they did not wish for India to go ahead and build a nuclear bomb. Bhabha continued to advocate the robust nuclear policy even after the passing away of Jawaharlal Nehru, and the ascension of Lal Bahadur Shastri to the position of Prime Minister. In fact, with Shastri's speech at the Lok Sabha meeting on 23-24 November 1965, Bhabha's position seemed to be validated. In his speech, Shastri said,

> Dr Bhabha has made it quite clear to me that as far as we can progress and improve upon nuclear devices, we should do so, as far as development is possible, we should resort to it so that we can reap its peaceful benefits and we can use it for the development of our nation...Just assume that we have to use big tunnels and we have to clear huge areas, we have to wipe out mountains for development parks, and in this context if it is required to use nuclear devices for the good of the country as well as for the good of the world, so then our Atomic Energy Commission is pursuing these same objectives.
>
> (Perkovich,1998:82)

Bhabha, however, died when an airplane carrying him crashed into the Mont Blanc on 24 January 1966. While his death was certainly a shock to all, his legacy lives on in the Atomic Energy Department of the country and beyond.

Assessment of Bhabha's contributions towards the nuclear program

While Homi Jehangir Bhabha will always be remembered as the progenitor of a program which has since been carried forward by many of his protégés, Bhabha has permanently found place in the annals of nuclear history. However, his methodology has been criticized on account of the excessive secrecy – given the fact that the program was ostensibly directed towards peaceful purposes and there was always the normative aversion to an overt weapons program.

However, this by no means undermines his contribution to the overall development of the program. He started the TIFR, masterminded the development of the AEC and eventually the DAE. The major criticisms levied against him were – an almost exclusive maintenance of control over the entire atomic structure, lofty promises, and above all, a rather non-democratic way of running his establishment. But, what one must understand is that the doings of Bhabha were not against what was required at the time. Decisions had to be taken quickly and with haste. It was not possible then to rely or depend on so-called procedures that would most likely lead to delays in decision making when efficiency was the order of the day. Today, for most people, Homi Bhabha remains a complex and contested figure. Some argue that he definitely is a person par excellence while others still believe that personal objectives drove him to do whatever he did in the name of national advancement. To conclude, we would do well to remember the following, which probably goes to sum him up:

> Since I cannot increase the content of life by increasing its duration, I will increase it by increasing its intensity...I do have this one purpose – increasing the intensity of my consciousness and life.
>
> (Homi Bhabha)

REFERENCES

- Venkataraman, G. *Bhabha and his Magnificent Obsessions*, Universities Press, 1994.
- Deshmukh, C. *Homi Jehangir Bhabha*, National Book Trust, 2010.
- Abraham, I. *The Making of the Indian Atomic Bomb: Science, Secret and the Postcolonial State*, London: Zed Books, 1998.
- Anderson, Robert S. *Nucleus and Nation: Scientists, International Networks, and Power in India*, University of Chicago Press, 2010.
- Perkovich, G. *India's Nuclear Bomb: The Impact on Global Proliferation*, University of California Press, 1999.

1. Bhabha and his parents 2. CV Raman (far right) and Homi Bhabha (far left) 3. Dr Bhabha (far right) with Professor Albert Einstein (far left) 4. Bhabha in front of the Cascade Generator, TIFR Archives

VIKRAM SARABHAI
An Institution Builder

Sameer Patil and Arun Vishwanathan

> There are some who question the relevance of space activities in a developing nation. To us, there is no ambiguity of purpose. We do not have the fantasy of competing with the economically advanced nations in the explorations of the moon or the planets or manned space flight. But we are convinced that if we are to play a meaningful role nationally, and in the community of nations, we must be second to none in the application of advanced technologies to the real problems of man and society, which we find in our country.
>
> – Vikram Sarabhai

GIVEN CONCERNING CONDITIONS of abject poverty, weak industries, and nascent political institutions, the allocation of post-Independence India's limited national resources was a highly debated issue. In this scenario, among the proponents of investment of national resources for scientific and technological advancement, Dr Vikram Sarabhai's was one of the strongest voices that rationalised technological advancement as a determinant of national progress. His perspective about scientific research rested on the fact that any invention, discovery, or innovation emerges from a real-world problem that it seeks to resolve – including impacting the lives of a nation's citizens as well as the standing of a nation at the international level. This idea of technological advancement led him to far surpass conventional boundaries and emerge as a visionary institution-builder par excellence. A physicist, an entrepreneur, an educationist, an art-appreciator, a leader, a visionary – Vikram Ambalal Sarabhai was a man who wore many feathers on his hat.

Early life

He was born on 12 August 1919 in pre-independent India's Ahmedabad, to a family of affluent industrialists. Born with a silver spoon, he developed an affinity towards science and innovation very early in his life. Along with his two brothers and five sisters, he completed his primary education from a private school started at their house, The Retreat, that was based on the Montessori School System. Thereon, having completed his matriculation, he enrolled in the Gujarat College (then affiliated to the Bombay University) in 1935 and completed his intermediate. In 1937, he moved to join St John's College at the Cambridge University, United Kingdom, for his undergraduate degree in Physics and Mathematics. As the World War II started in 1939, he returned to India and continued his post-graduate education at the Indian Institute of Science (IISc), Bangalore, under the guidance of Nobel Laureate Sir CV Raman. It was Professor Raman who encouraged Sarabhai to undertake his post-graduate and doctoral studies in Cosmic Rays. His first research paper titled 'Time Distribution of Cosmic Rays', published in *Proceedings of Indian Academy of Sciences* in 1942, dealt with the probability distribution of α-particles emitted during radioactive decay (Joshi, 2019).

Sarabhai carried out his post-graduate research experiments across three cities – in Bengaluru, in Pune and in Apharwat (a place in the Himalayan mountains in the Kashmir Valley). It is during these experiments that he envisioned the establishment of a dedicated laboratory for studying cosmic rays and atmospheric physics even at this early stage of his research. His association with Dr KR Ramanathan at the Poona Observatory at the Indian Meteorological Department (IMD) further drove this idea to be materialised in Ahmedabad. He started the Physical Research Laboratory (PRL) in 1945, in association with Karmakshetra Educational Foundation, that was started by his parents. Moreover, the same year, as World War II came to a close, he moved back to Cambridge University to complete his PhD, this time with his wife in company. He invited Dr Ramanathan to join as the Director of this proposed new laboratory. Dr Ramanathan accepted and joined in 1948 post his retirement from IMD (Joshi, 2019).

Sarabhai continued researching on photo-fission for his doctoral thesis titled 'Cosmic Ray Investigations in Tropical Latitudes' for Cavendish laboratory until 1947, following which he returned to India. By 1957, as the International Geophysical Year Program started, Sarabhai realised the dearth of institutes in India pertaining to any research regarding the outer space. Cognizant of the potential of the field and its significance in the future, he proposed to the Government of India to commence an Indian space programme (Bhatia, 1985).

The nuclear programme under Sarabhai

The foundations for the nuclear programme of India were laid by Dr Homi Jehangir Bhabha, who led the mission in an unconventional manner that was way ahead of his times. However, Dr Bhabha's untimely death in an Air India plane crash in January 1966 came as a great national loss as far as nuclear technological advancement was concerned. Just before his death, Bhabha had appointed Sarabhai as a member of the Atomic Energy Commission (AEC); this brought Dr Sarabhai at the helm of affairs of the Indian nuclear programme. However, unlike Bhabha, Sarabhai's position was not *numero uno*. The positions that Bhabha occupied were divided among three people after his death. Sarabhai headed the Atomic Energy Commission (AEC) and the Department of Atomic Energy (DAE), whereas one of his colleagues, Homi Sethna, was appointed Director of the Atomic Energy Establishment (AEE), Trombay (soon renamed as the Bhabha Atomic Research Centre or BARC). Meanwhile, another scientist MGK Menon was chosen to head the Tata Institute of Fundamental Research (TIFR). In addition to this, under Prime Minister Indira Gandhi, the mutually supportive relationship between the Prime Minister and the head of the nuclear programme (as seen earlier between Jawaharlal Nehru and Dr Bhabha) was revived. However, Indira Gandhi and Dr Sarabhai shared an antipathy towards the atomic bomb, and in that sense, Sarabhai's position was a departure from his predecessor. He is said to have deprioritised the requirement of the Peaceful Nuclear Explosion owing to his Gandhian-pacifist leanings. He is in fact believed to have kept to himself proposals formed under Bhabha, never bringing it out for discussion after Bhabha's death.

Having said that, his contribution to defence technology cannot be ignored. During the Indo–Bangladesh war of 1971, when short-range solid fuelled rockets, which could send more explosives across the border, were required, it was Sarabhai who sanctioned them (Prahladan, 2016). After Sarabhai's sudden death in 1971, Homi Sethna succeeded him and went on to play a major role in Operation Smiling Buddha, the Peaceful Nuclear Explosion in May 1974.

Despite Sarabhai's contribution to the nuclear programme, he was still considered an outsider by the rest of the scientific nuclear establishment. As a matter of fact, some of his peers even questioned his credentials to lead the atomic energy programme, casting doubt on his knowledge of the subject and experience. In this context, Ashok Parthasarathi writes that Dr Sarabhai's 'legitimacy' to lead the programme was questioned by the BARC community right from the time he became the AEC Chairman. The argument was that 'he did not have the knowledge base and experience in nuclear science and engineering that Sethna or Ramanna had' (Parthasarathi, 2001). Expectedly, Homi Sethna was one of the leading scientists opposing Sarabhai, as the former had expected to succeed after Bhabha's death, and towards that end reportedly produced 'a letter from Bhabha saying he would be in charge in Bhabha's absence' (Anderson, 2010). However, when he did not get the job, Sethna adopted a 'bitterly adversarial position' till Sarabhai was alive – what Raja Ramanna, who became the head of the nuclear programme in the 1980s termed as 'jealous colleagues' (Ramanna, 1991). In fact, another example of disagreement within the DAE – between Dr Sarabhai and Homi Sethna, on the Non-Proliferation Treaty (NPT) – reinforces this point. At a time when Sarabhai had been spelling out to the United States and other nuclear powers India's position on disarmament and security guarantees, along with the opposition to the NPT, Sarabhai's colleague from the Atomic Energy Commission, Homi Sethna had a different point of view. Speaking to the American Embassy officials, he had remarked that for him the 'only' thing objectionable in the NPT was the 'inspection provisions' (Department of State, 1968b).

In order to deal with this internal opposition, Dr Sarabhai involved the existing top leadership in the atomic energy department in policy decision making, thereby attempting to win them over and alleviate their concerns about the manner in which he would take the nascent department forward after the loss of such a tall leader as Homi Bhabha. During his tenure, Sarabhai made an attempt to move the AEC's policy-making mechanism towards a 'regulatory' system even though top policy decisions mainly remained in the domain of the Prime Minister. True to his collegial style of functioning, Sarabhai sought to accommodate his eminent colleagues in all aspects of nuclear decision-making, including Homi Sethna and Raja Ramanna (Parthasarathi, 2001). In terms of decision making and management capabilities, Dr Ramanna has been quoted in Padmanabh Joshi's excellent book on Sarabhai as stating, 'I had some doubts as to whether Vikram would be able to get a grip on atomic energy developments here and abroad, and be able to give it special orientation. But in only a few months, he was able to participate in the deliberations of all the decision-making bodies in an effective and constructive way (Joshi, 1992).' Yet, in the end, he largely failed to secure the backing of his colleagues, particularly those from the BARC (Parthasarathi, 2001).

Internal oppositions aside, it is noteworthy that Sarabhai's contribution to India's nuclear programme cannot be underestimated. After Bhabha, it is Sarabhai who deserves credit for the establishment and development of nuclear power plants in the country. Moreover, as mentioned before, his push towards self-reliance led to laying the foundation of indigenous development of nuclear technology for defence purposes.

Sarabhai: The father of India's space programme

However, when it came to the space programme, Sarabhai enjoyed the same stature that Bhabha enjoyed in the nuclear programme. Regarded as the 'father of the Indian space programme,' Sarabhai had set up the PRL in Ahmedabad in

1947, in his mid-twenties. Just like its counterpart in nuclear research (the TIFR, set up two years earlier), the PRL came to be known as the cradle of the Indian space programme. Later, in 1962, Indian National Commission for Space Research (INCOSPAR) was set up by the Government of India under the DAE, with Sarabhai appointed as the Chairman. He was additionally in charge of the administrative responsibilities of INCOSPAR. The PRL, by implication, acted like the headquarters of INCOSPAR. The INCOSPAR was later renamed as the Indian Space Research Organisation (ISRO). The following year, at the United Nations, he proposed the establishment of an International Equatorial Sounding Rocket Launching Facility in the country. The proposal was accepted and it was established in Thumba, Thiruvananthapuram. The site, due to its proximity to the geo-magnetic equator, was the most appropriate place for India's first space centre. In November, the same year, the country's first rocket called Nike-Apache, was launched from the Thumba Equatorial Rocket Launching Station (TERLS). TERLS was formally dedicated to the United Nations in 1968 by then Prime Minister Indira Gandhi (ISRO, 2018).

Along with AEC and INCOSPAR, Sarabhai had been additionally shouldering the responsibility of the Electronics Committee. Knowing his position fully well, when he became the head of the DAE, Sarabhai sought to fuse the space and nuclear programmes – as two flagships of India's scientific advancement (Bhatia, 1985). The Indian Space Research Organisation (ISRO) was subsequently established in 1969, during his tenure. In terms of prioritization of fields, the space programme under Sarabhai developed a targeted focus at building India's satellite technology. In this pursuit, in 1970, he persuaded the government to approve a ten-year plan towards achieving self-reliance in space research (Bhatia, 1985).

Apart from managing and driving the progress of India's space programme from within, he undertook a number of international missions. As the Vice-President and the Scientific Chairman of the UN Conference on Exploration and Peaceful Uses of Outer Space, held at Vienna, during August 1968, Sarabhai made a case for the necessity of developing and underdeveloped nations to adopt scientific advancement and progress as a priority issue, just like poverty and malnourishment. He believed that scientific advancement was an important tool for reduction of inequality between the developed and the developing world. He said,

> A developing nation following a step-by-step approach towards progress is landed with units of small size, which do not permit the economic deployment of new technologies. Through undertaking ventures of uneconomic size with obsolete technologies, the race with advanced nations is lost before it is started. Indeed, if one continues to operate on this philosophy, financial and technical assistance from the advanced nations to the developing nations can only result in the frustration of the former and the increasing economic dependence of the latter.
>
> (Joshi, 2019)

Sarabhai's multifaceted contributions towards national progress

The excellence of Sarabhai as an institution builder was such that his contributions to national progress and development went way beyond the space and nuclear programmes of the country. His dynamism and drive for institution building can be attested by the fact that between 1948 and till his death in December 1971, he is believed to have established one institution per year on an average (Ganesh and Joshi, 1985). Apart from PRL, he established the Ahmedabad Textile Industry's Research Association (ATIRA) in 1947. He managed its affairs until 1956. Identifying the requirement of professional management education in India, he played a crucial role in establishment of the Indian Institute of Management in Ahmedabad (IIM-A) in 1962 – the same year when the foundation of INCOSPAR was laid (*Who Was Vikram Sarabhai*, n.d.). Additionally, in line with his idea of scientific advancement aimed at solving real world issues, he was a prominent proponent of ensuring accessibility of education to remote areas through satellite communication. According to McCracken, he proposed having a single video channel, fourteen different sound channels in different languages to ensure that linguistic differences are not a hurdle (Pal, 2013). He further proposed the use of satellite-based remote sensing of natural resources. He claimed that using a satellite to deliver a high-quality countrywide telephone system would be faster than using a traditional ground-based microwave system. That is to say, he spoke of carefully fitting space science to his country's national aims (Pal, 2013).

Additionally, he set up India's first market research organisation, the Operations Research Group, far before the significance of market-based research was properly recognised and appreciated. With his wife, Mrinalini Sarabhai, a world renowned classical dancer, he set up the Darpana Academy of Performing Arts in Ahmedabad. As Sarabhai's biographer Amrita Shah recalls, he was supportive of Mrinalini's professional career, even taking care of the child when she travelled abroad (Shah, 2022). Yet, she also adds that Sarabhai's extra-marital affair with another woman devastated Mrinalini. She couldn't reconcile with it till Sarabhai was alive (Shah, 2007).

Other major institutions set up by him include the Space Applications Centre, Ahmedabad (formed by merging six institutions established by Sarabhai), Faster Breeder Test Reactor in Kalpakkam, Variable Energy Cyclotron Centre in Kolkata, Electronics Corporation of India Limited in Hyderabad, Uranium Corporation of India Limited in Jaduguda and Community Science Centre in Ahmedabad.

Untimely demise

Dr Vikram Sarabhai was taken from our midst on 30 December 1971 at the very young age of fifty-one. That in such a short time, he had touched so many people, institutions and areas is something extremely rare. Dr Sarabhai was a man of multiple talents having vast and in-depth knowledge on a variety of subjects. In a very poignant obituary published in *Nature*, Dr Sarabhai's life is described in the following words:

> … In his style of working, 'hurry' had a charming leisurely

quality about it, even though every minute of his time was claimed by one or the other of his self-created projects. Within the space of an hour, he could go from cosmic-ray modulation, to rocket propellants at Thumba, to power reactors at Kalpakkam, to the television satellite experiment, to community science education, to antenna systems for communication links, to infrared devices for surveillance of crop disease, to the administrative details of setting up a new public corporation and back perhaps to cosmic rays. And each of these transitions was deep, involving decision making, unwasteful of words, and outwardly done with a naturalness and charm which always overwhelmed his friends. Vikram Sarabhai will be missed on the Indian scene, as a scientist, as a scientific visionary, as a doer of things and as a remarkable human being.

(*Dr Vikram A Sarabhai*, by H Dale, *Nature* Vol 236, March 1972).

Professor CNR Rao also recalls Sarabhai as a workaholic. As he notes in his obituary,

Vikram Sarabhai used to work very hard. Whatever he took on, got his complete attention and that was so with respect to the space programme as well. He used to go back and forth between Bombay and Trivandrum, and day or night did not matter to him. I still remember Professor Raman advising him to be careful, not to travel too much, and take care of his health. I used to see him travelling on planes going through innumerable files during the trip.

(Rao, 2020)

Such was the pace of his working style that his colleagues were simply unable to keep up with him, leading his biographer Amrita Shah to conclude that 'Vikram tended to have a somewhat unrealistic perception of ordinary human capacities' (Shah, 2007).

The visionary's legacy

The visionary scientist and leader has left behind a glorious legacy. He was awarded the Shanti Swarup Bhatnagar Prize in 1962 by the Council of Scientific and Industrial Research. Later, he became the recipient of two of the highest civilian awards of the country, namely the Padma Bhushan in 1966 and the Padma Vibhushan posthumously in 1972. Pramod Kale writing in the Special issue of *Current Science* commemorating the centenary of Vikram Sarabhai's birth, points to the manner in which Dr Sarabhai would mentor young students, scientists and technical staff during his stint at the PRL. Kale writes that Dr Sarabhai would often drop by unannounced at the PRL research laboratory tracking satellites (Dr Vikram A Sarabhai, 1972). This trait of Dr Sarabhai has been described by many people who have worked with him. Reiterating this opinion, SR Ganesh and Padmanabh Joshi too have encapsulated the broad guiding principles with which Dr Sarabhai used to build institutions and mentor the next generation of people within those institutions. They state that Dr Sarabhai's guiding principles – whether it be ATIRA, PRL, ISRO or other multiple corporate or educational institutions – can be defined as a set of three fundamental principles:

- Networking strategy or creating interacting and overlapping clusters internally as well as externally both to produce a vision for the institution and to translate the vision into actions in terms of research programmes and projects;
- Trusting strategy or creating a climate of trust providing freedom of action to the individuals, ensuring autonomy, and emphasizing horizontal control; and
- Caring strategy or creating a climate of caring by the leader remaining approachable through open channels of communication and emphasizing the role of administration as a support system to the core tasks of the institution.

(Ganesh and Joshi, 1985)

His memory has been kept alive through a number of symbolic and other gestures offering him his due credit for his contributions to India's growth story as well as to the international scientific community. In 1973, the International Astronomical Union in Sydney, Australia, named a crater on the moon in Sarabhai's name. ISRO recently launched the first lander-rover module from India to fly to the moon, to investigate it, on 22 July 2019. Vikram was the name of the lander that carried the rover. In his honour, ISRO's principal facility for launch vehicle development, located in Thiruvananthapuram, has been named the Vikram Sarabhai Space Centre (VSSC). On the first anniversary of his death (30 December 1972), the Indian Postal Department issued a commemorative postal stamp.

However, the biggest weakness of Sarabhai, like Bhabha,

was a failure to 'try to build into the structure and practices of the ISRO an institutionalisation of leadership capabilities, which would ensure that the next generation leadership was identified and groomed well in time' (Parthasarathi, 2001). Sarabhai's failure to raise and mentor the next set of leaders raised questions on his leadership strategies (Parthasarathi, 2001). As Parthasarathi rightly noted;

> When one looks back at it now, one cannot but be amazed at the inappropriateness and impracticality of Sarabhai's actual leadership style when it was evident, even as early as 1967-68, that ISRO would become a major agency demanding a plateau-like rather than a pyramidal-type leadership style. Rather than instituting an all-consultative mechanism to ensure a whole-of-ecosystem approach, Sarabhai hardly consulted his colleagues on matters of space. This was in contrast to his style in the nuclear domain, where, as noted earlier, he had enforced a collegial style of functioning involving his colleagues. This was evident when Sarabhai presented to the AEC, then the cabinet and finally the country in 1970, a '10-year profile' on both atomic energy and space drawn up with hardly any inputs from DAE and only limited inputs from ISRO.
>
> (Parthasarathi, 2001)

Conclusion

In conclusion, it may be said that the contributions of Vikram Sarabhai towards scientific development in the country have been unparalleled. These contributions touched a million lives and continue to be a guiding light for national progress till date. Ideas initiated by him decades ago have become an essential part of citizens' lives in the contemporary world – be it distance education, satellite communication, market-based research or remote sensing of natural resources. Being a man who thought way ahead of his times, Vikram Sarabhai indeed laid a very strong foundation of India's nuclear and space programmes that multi-dimensionally form an essential component of India's growth story – leaving behind a glorious legacy to carry forward.

REFERENCES

- Anderson, R.S. *Nucleus and Nation: Scientists, International Networks, and Power in India.* University of Chicago Press, 2010.
- Bhatia, A. India's Space Program: Cause for Concern. *Asian Survey, 25*(10), 1013–1030, URL:http://www.jstor.org/stable/2644178, 1985.
- Department of State, Telegram: Conversation with Senior GOI Nuclear Official. Secret. 7 May 1968, The National Security Archive of the George Washington University, [Online: web] URL:https://nsarchive2.gwu.edu/NSAEBB/NSAEBB6/docs/doc17.pdf, 1968b.

- Ganesh, S. R., and Joshi, P. Institution building: Lessons from Vikram Sarabhai's leadership. *Vikalpa, 10*(4): 399–413, URL:https://www.isro.gov.in/sites/default/files/flipping_book/institution_building/files/assets/common/downloads/Institution%20Building.pdf, 1985.
- ISRO. *Golden Jubilee Celebration of TERLS*, https://www.isro.gov.in/golden-jubilee-celebration-of-terls#:%7E:text=On%20February%2002%2C%201968%2C%20the,international%20range%20for%20carrying%20out, 2018.
- Joshi, P. *Vikram Sarabhai: The Man and the Vision*. Mapin, 1992.
- Joshi, P. Dr. Vikram Sarabhai: Father of India's Space Program. *Physics News, 51*(37): 4–7, URL:https://www.tifr.res.in/~ipa1970/news/2019/A1_PN_PJoshi_fn.pdf, 2019.
- Paikowsky, D. *The Power of Space Club.* Cambridge University Press, 2017.
- Pal, Y, *Space and Development: Proceedings of the Vikram Sarabhai Symposium on Space and Development of the Twenty-second Plenary Meeting of COSPAR Bangalore, India, 29 May to 9 June 1979*, Pergamon Press, 2013.
- Parthasarathi, A. Leadership in Science and Technology: Some Aspects of the Indian Experience. *Economic and Political Weekly, 36*(40): 3843–3851, URL:http://www.jstor.org/stable/4411204, 2001.
- Prahladan, V. 'Indira Gandhi and the nuclear bomb: She never gave the green signal', Living Media India Limited, URL: https://www.dailyo.in/politics/indira-gandhi-nuclear-bomb-homi-bhabha-pne-vikram-sarabhai-pokhran-test-barc-aec/

story/1/14163.html, 2016, November 22.

- Ramanna, R. *Years of Pilgrimage, an Autobiography*, Viking, 1991.
- Rao, C.N.R. Tribute to Vikram Sarabhai, *Current Science*, 118 (8): 1182, 2020.
- Sarabhai, V. A. Ahmadabad Textile Industry's Research Association, *Nature*, *174*:578–580, URL:https://www.nature.com/articles/174578a0, 1954.
- Shah, A. *Vikram Sarabhai: A Life*, Penguin, 2007.
- Shah, A. '"Rocket Boys" gets Vikram Sarabhai wrong', *Indian Express*, URL:https://indianexpress.com/article/opinion/columns/rocket-boys-vikram-sarabhai-sony-liv-7796604/, 2022.
- Vikram A. Sarabhai, *Nature*, *236*(5343):185–186, URL: https://doi.org/10.1038/236185a0, 1972.
- 'Who was Vikram Sarabhai' (n.d.). *Business Standard*, URL:https://www.business-standard.com/about/who-is-vikram-sarabhai.

1. Dr Sarabhai with his wife and son 2. Childhood picture of Sarabhai after he built an engine 3. Noble Laureate CV Raman (far right) and Vikram Sarabhai (third from left) at the CSC's foundation stone laying ceremony in Ahmedabad 4. Dr Vikram Sarabhai with his wife and children

RAJA RAMANNA
Nuclear Physicist with a Piano

Shrabana Barua

RAJA RAMANNA HAS been one of the most important personalities to contribute to India's nuclear and scientific endeavours throughout its nascent years and beyond. He has had great success as a scientist, musician, politician and as a human being, who has left a conspicuous mark on the Indian nuclear page that is hard to ignore. From being associated with the Tata Institute of Fundamental Research (TIFR) to heading the Study of Nuclear Explosions for Peaceful Purposes (SNEPP) as a director of physics at the Atomic Energy Establishment (AEE), he has gone on to lead Bhabha Atomic Research Centre (BARC) as a Director and the Atomic Energy Commission (AEC) as Chairman in an illustrious career spanning over four decades. He has engineered achievements of not just the popularly known Pokhran-1, but also made the Apsara, Dhruva, Fast Breeder Test Reactor (FBTR) and use of the atom for various purposes a success in India. The multiple honours bestowed on him were well deserved and his emphasis on self-sufficiency for India's development in the nuclear field and otherwise is of immense relevance. It is challenging to jot down the life and contribution of such a vibrant person within the scope of this chapter. However, an attempt has been made to do the utmost.

Childhood and student life

Rukmini Amma was a 'voracious reader' (Ramanna, 1991:22) with a progressive outlook, while B Ramanna was a judge in Mysore. They married at the early age of thirteen and had six children. Raja Ramanna was the youngest of them.

Born in Tumkur, Karnataka on 28 January 1925, Bindiganavile Krishnaraja Ramanna was named after his maternal great grandfather Bindiganavile Krishnappa popularly known as Krishna Iyengar, a revered crusader of Mysore's identity in the nineteenth century, and his aunt Rajamma, who had influenced young Ramanna in many ways. Raja Ramanna cut short his name, as he liked it so. As a child, he lived in Mysore and studied in Dalvoy School. At the age of six, he shifted to Good Shepherd Convent near Bangalore where he started developing his association with western music. Ramanna's admiration for the Hungarian composer Franz Liszt was no secret and had great influence on him as a musician. By 1935, the family shifted to Bangalore where Ramanna joined the Bishop Cotton School. In Mysore, he had a taste of courtly grandeur at the age of twelve when Maharaja Krishnaraja Wadiyar IV took note of the pianist in him and invited him to join the orchestra at Jaganmohan Palace in 1937, not once but twice! He even received a scholarship to learn music in Trinity College that year, which he could not avail however.

Raja Ramanna was a privileged child and student. His kith and kin were influential in their own capacities, at least in Mysore. His close association with the Palace further gave him a sense of security; it also made him attuned to people's behaviour around circles of power. His experiences as a child, his trips to various places in India moulded him into a person with a positive outlook and vision about things in general.

After completing his intermediate studies from St Joseph, Bangalore, Raja Ramanna joined the Madras Christian College in Tambaram pursuing his favourite subject – Physics. During his BSc (Hons), he also continued to learn music, re-associating himself with the Mysore Palace band. It was during one orchestra rehearsal under Dr Alfred Mistowski in 1944 that he was first introduced to Homi Bhabha. Later Bhabha arranged the JN Tata Scholarship for Ramanna to pursue his doctorate at King's College, London. By September 1945, Ramanna set off aboard the ship *Orion* to United Kingdom (Ibid: 37).

By then, the nuclear bombs had just brought an end to the World War, and in fact a crater (symbolic of the war destruction) stood at the entrance of King's College. Ramanna confessed, 'the atom bomb's technological aspects made a deep impression on me which eventually led me to choose nuclear fission for my doctorate' (Ibid: 60).

He further discussed cosmic ray phenomenon and ionospheric

activity with his professors. Faced with discrimination and disappointment at being asked to enrol for MSc (despite having the requisites to register for PhD), he found some respite under the guidance of Dr Alan Nunn May. Soon, however, Dr May was arrested for 'leaking atomic secrets to the Russians' and Ramanna was back under the supervision of Dr FC Champion (Ibid: 39-40). Somewhere, this uneasy experience, back in the so called 'advanced colonizing nation' of Britain, was perhaps the reason behind Ramanna's deep quest for self-sufficiency for his own country throughout his life.

Ramanna completed his PhD in 1948 and was also 'offered post(s) at Dr Bhabha's laboratory' (Ibid: 51). Meanwhile in London, he had also registered at the Trinity College of Music to continue learning piano. As Bhabha's efforts to create the TIFR had borne fruits by then, he wanted Ramanna to stay on with an AEC scholarship to further delve into the area of fission problem. It was on 1 December 1949 that (by then Doctor) Raja Ramanna returned to India with a hope for a role in contributing to the progress of the country. In his book, *The Years of Pilgrimage: An Autobiography*, Dr Ramanna writes:

> My time as a student is something I look back upon with great fondness, for despite the tensions and disappointments, my student days were essentially exciting and carefree; a period when without quite knowing it, we were on the point of bidding farewell to immaturity and adolescence and entering the world of responsible adulthood.
>
> (Ibid.)

Evolution as a scientist

Back in London, Ramanna had managed to build an ionising chamber which he brought back with him. Besides his thesis, he had gained experience in the field of nuclear fuel cycles and reactor designing as part of a research at the Atomic Energy Research Establishment (AERE). He was quite ready to take on the new role he would be entrusted with and evolve into the great scientist that we know of today.

Work at TIFR

On his return, Ramanna was absorbed into the TIFR in the first week of December 1949 itself. He was full of praise (like most others) for Homi Bhabha's efforts at the institute as 'they became the forerunners of all indigenous technological activity in the country and heralded the beginnings of an extensive atomic energy programme in India' (Ibid: 56). Ramanna believed that despite lag in technical and technological aspects in India and insufficient assistance for scientific efforts at that time, commendable efforts brought fame to the TIFR, mainly due to Bhabha's vision. If so, the fact that India insisted on self-sufficiency for itself in the field of nuclear technology, building all parts and processes indigenously, has to be attributed much to Dr Ramanna's zeal. Interestingly, even Bhabha was full of consideration for Raja Ramanna and allotted him two adjacent rooms at TIFR – one for the scientist and the other for his piano (Ramanna, 2004:1150-51)!

By this time, TIFR had various branches led by different people. Ramanna worked from the ground floor of the building which was the laboratory of nuclear physics. It was in vogue to study particle physics and quantum mechanics. However, instead of following the crowd, Dr Ramanna chose to concentrate and find solutions to the 'many-body problem' of nuclear fission and focussed on research in modern technology like semi-conductors, lasers, superconductors, etcetera with the aim of making India take long strides in comparison to the world at large. Ramanna in his autobiography wrote that the AEC created in 1948 was 'not only for the development of nuclear power, but was given the clear directive that its goal should be self-sufficiency' (Ramanna, 1991:60). Indeed, Ramanna took this very seriously. His work on areas of cosmic ray and nuclear scattering, role of neutrons, nuclear and reactor physics placed him among pioneers in the field in India.

Association with BARC

In 1952, Dr Ramanna married Malathi Iyengar, daughter of K Narasimha Iyengar, and eventually had three children, two daughters and a son. On the work front, he joined the Atomic Energy Establishment, Trombay in 1954 (which in 1967 was formally named as BARC).

The Department of Atomic Energy was established in mid-1954 and had great powers. By 1955, what Itty Abraham calls the 'nuclear enclave' had come under immense criticism for its fetish for control over the nuclear programme and secrecy structured by Bhabha (Abraham, 1999). The national symposium organized in 1955 saw Meghnad Saha vigorously attempting to expose some pitfalls of AEC. It is said:

> The most tangible challenge the critics were able to lay down was to suggest that the AEC, as currently constituted, would not be able to build a working nuclear reactor for many years to come.
>
> (Ibid: 76)

Being bound by the goal of self-reliance, it was clear that a reactor could not be imported altogether at once, and yet it became imperative to seek foreign assistance. It was then that Bhabha arranged for fabricating the components (as suggested by UK) to build a small swimming pool research reactor in 1955. This work was done by a team consisting of AS Rao, NS Prasad, HN Sethna, KS Singhvi and Raja Ramanna. Overcoming various difficulties of leakages, heavy monsoon, doubts and tensions, the group of scientists worked relentlessly to give, within a year, India its first indigenous research reactor on 4 August, 1956 – the Apsara reactor. It enabled India to have its own reactor for conducting scientific experiments and enhancing its know-how in the field of nuclear and reactor physics, something which Ramanna had already been working on. He began conducting other experiments here onwards which helped him to explain stochastic theory later (for which he felt he wasn't rightly acknowledged however). Dr Ramanna stands out within this group for being the youngest scientist involved in the project. This achievement made the 'Indian nuclear energy establishment a force to reckon with' (Ramanna, 1999:63). Indeed 'the first taste of success remains unparalleled' but Dr Ramanna had many more to come.

Another area where Ramanna's role was crucial was in neutron production in India. Later, Ramanna would enumerate the various uses of isotopes which are produced by neutrons and reiterate at various occasions the peaceful uses of the atom in the new century. At that point, it was Ramanna who convinced Dr Bhabha in 1955 (few months before the Geneva Conference on the Peaceful Uses of Atomic Energy) to import 'a cascade generator for neutron production in India' (Ibid.). Due to his insistence, India even received processed beryllium oxide from France. This helped to carry out experiments to measure value of neutron diffusion using beryllium oxide (an efficient moderator) through pulse technique, a contribution which India in fact made at the Geneva Conference that year and earned praises for. Ramanna wrote with pride:

> The paper published by us on the subject came as a surprise to the French. They had been working on it for a long time, trying to measure the same parameters by a conventional method which required tons of beryllium. In comparison, the pulsed neutron techniques used by us required only a few kilograms which cut down the cost as well [as] the time-span for carrying out the experiment. An interesting device for measuring time was the time analyser that we had developed.
>
> (Ibid.)

As a student, though nobody speculates much on the teaching methodology of schools or colleges, in retrospect, many a time, one wonders what impact such teaching has had on oneself. Raja Ramanna too pondered on the education and training one receives as a student in India, and whether it helps to solve issues of real life. For his own part, he felt that, back as a student in Madras Christian College, most of his physics was taught from an exam point of view without the student being involved in any other way. Perhaps, with some hope of changing such a predicament, Dr Ramanna facilitated the creation of the BARC Training School in 1957 to enable training in nuclear science in India and enhance skilled manpower for dealing with multiple problems of a nuclear programme in practicality. With the aim of nurturing the future generation in the field, he also believed it 'helped greatly in stalling the emigration syndrome' (Ibid: 79). M Srinivasan later wrote:

> The legacy of Ramanna is that over a half century of his association with [the] atomic energy programme, he helped build up a large pool of scientists and technologists who could take on new and challenging problems in nuclear science and technology for national progress...
>
> (2002)

By 1958, Dr Ramanna had become the Chief Directing Officer at the AEE. He began to shape the Indian nuclear programme in ways that were subtle yet important. In the 1960s, the CIRUS (Canada India Research Utility Services) reactor was built, CANDU RAPS 1 (Canada-Deuterium-Uranium Rajasthan Atomic Power Station) reactor was in place without Canadian strings attached, and Tarapur power station was in progress with the US fuel supply to enable India nuclear power generation from nuclear energy (a concept well propagated by Ramanna in the 1980s). While India negotiated for assistance, the need was felt to build a self-reliant nuclear programme for India. The safeguards of the International Atomic Energy Agency (IAEA) were now in operation as well. Raja Ramanna and others knew that making use of the immense thorium reserves in the country was the only way to realize the dream of self-sufficiency. Yet, plutonium had to be produced which could then be integrated into the fast breeder reactor scheme. Options were being considered on the suitability of importing reactor technology. It was with clear sightedness and resoluteness that Dr Ramanna and his team tried to convince Bhabha to consider the option of heavy water natural uranium reactor as best suited for the time. Despite doubts about the heavy water reactor (which did not work well initially), Dr Ramanna adhered to his views; and in a speech during the Meghnad Saha Memorial Lecture in 1984, he said;

> [T]he far sighted decision over two decades back, to opt for PHWR instead of BWR has probably contributed most to the self-reliance which India has acquired in the field of nuclear power generation.

He further believed it would enable India its 'nuclear option' and help keep away these reactors from inspection clauses. If Ramanna's report on the internal discussions regarding the implications of reactor choice in 1954 is put together, it becomes quite evident that it was with the intention of keeping reactor and fuel away from safeguards (that would eventually enable a weapon option) that the CIRUS negotiations were made (Abraham, 2001:122). Given that India did not have any weapon programme in mind, it could be questioned as to why such a move was at all needed. The answer lies, however, in the ever pertinent goal to keep away interferences (in the name of safeguards) of the advanced nations. Ramanna's view of the Atoms for Peace Plan of 1953 was sanguine, but he had figured out that it was in the commercial interest of the enriched uranium sellers to project the virtues of atomic energy for electricity generation and then keep a check on proliferation or even mere nuclear activity in the buyer nations. By mid-1960s, the advanced nations had begun deliberating on the Non-Proliferation Treaty (NPT) as a way to cap nuclear activity in the not-so-advanced states. In any case, India managed to make its own natural uranium fuel (with the spent fuel of the CIRUS reactor) that entered the plutonium reprocessing Phoenix plant in 1964 (Perkovich,1999:64). It gave India a major boost in technological progress, and as Ramanna notes, AEE 'became one of the foremost laboratories of the country' (1991:71).

Dr Ramanna participated in IAEA programmes and proposed solutions for bringing about regional cooperation among Southeast Asian nations for optimum utilization of West-provided research reactors. He advocated the usefulness of research reactors in India for peaceful purposes. There is little doubt that this remained his main purpose throughout his life. During a short interaction with this author in 2015, Dr George Perkovich, who had interviewed Raja Ramanna twice while working on his seminal book *India's Nuclear Bomb: The Impact on Global Proliferation*, fondly recalled that Raja

Ramanna was a 'very decent man' who never advocated the bomb for the sake of it despite being a part of its creation (at a seminar in CIPOD, JNU on 3 November 2015). Indeed, later Ramanna wrote about his desire to join in any movement led by USA to ban nuclear weapons and even 'plead[ed] for that movement'.[1]

Meanwhile his efforts were acknowledged by Jawaharlal Nehru, who expressed the view that 'India should give more support to science as the country's future' (as also its neighbours') rested more on science than on any other field of activity (Ramanna, 1991:73). Ramanna's work had begun to be recognised. The evolution of the scientist in him was in full bloom when he became the recipient of the Shanti Swarup Bhatnagar Award in the year 1963.

The milieu of India's nuclear programme had begun to change with Bhabha's 24 October 1964 All India Radio speech claiming that 'a 10 kiloton explosion…would cost Rs 17.5 lakhs' (Perkovich,1999:67). Ramanna would later admit that he didn't think it 'would be possible to do what Bhabha said – build a device in eighteen months' though an extremely expensive crash programme could increase its probability (Ibid., p.71). Anyhow, by 1966, AEE had permission to theoretically begin preparations for a Peaceful Nuclear Explosion, for though then Prime Minister Shastri was not aligned with Bhabha's aims, they began to converge at some point by late 1964. Following

1 Ramanna, R. 'Technology Transfer and Non-Proliferation', *Journal of Indian Ocean Studies*, NIAS.

Shastri's Trombay visit in 1965, Dr Raja Ramanna was officially nominated to head the SNEPP. He further became the Director of Physics at BARC in 1966. It was in this capacity that new and more prominent responsibilities awaited him.

Shastri and Bhabha's deaths within a span of few weeks in early 1966 made the Indian nuclear scene very murky. Indira Gandhi (less Gandhian than Shastri) became the Prime Minister. On the other hand, the AEC came under the Chairmanship of Vikram Sarabhai, a person with 'fundamental distaste towards nuclear weapons' who thought of it as a 'paper tiger'. Ramanna and Sarabhai were rarely on the same page. The new Chairman formally halted work on Peaceful Nuclear Explosions (PNE). Ramanna and his colleagues were upset at this, but seemed quite unmoved as experiments and research continued behind the scenes, despite the Chairman's awareness of it. He knew Sarabhai could not stop them from doing their work. In the following years, Ramanna quietly engaged the services of Dr Rajagopala Chidambaram to 'derive the equation of state of plutonium, a study necessary to design a nuclear explosive with plutonium core' (Perkovich,1999: 40). Thus, the seed for the 1974 PNE was sown under the leadership of Dr Ramanna by 1967 itself. The following year Dr Raja Ramanna received the Padma Shri Award.

In an interview with journalist Rajiv Mehrotra in 1990s, Ramanna was once asked what the future scientific research in India should aim for. He was clear in answering that scientists should aim for 'perfection' and not be 'satisfied with simple achievements', as was often the case in India (TV show

'In Conversation' hosted by Rajiv Mehrotra, and available on YouTube). This somewhat defined the character of Dr Ramanna as a nuclear scientist. He never gave up on the desire for India to conduct a PNE, in fact he pushed for a second test later with the aim to better the scientific achievements.

By 1971, Ramanna's team with PK Iyenger, R Chidambaram and others had the basic design of the explosive device ready, though associated devices and production of the plutonium alloy was only considered when time was more favourable. Even the Purnima reactor had been constructed which achieved criticality in 1972. Sarabhai's death in 1971 (with Homi Sethna replacing him) and promotion of Ramanna to the Directorship of BARC in 1972 catalysed the PNE project. Dr BD Nagchaudhuri of Defence Research and Development Organisation (DRDO) collaborated with BARC and the work was in full swing. The Prime Minister was now kept informed about the developments though not more than seventy-five scientists and engineers were involved directly in the project. The neutron trigger took some time, but all material problems had been tackled by 1973. By September that year, the green signal was received and site inspections and preparations for the 18 May 1974 PNE had begun.

In his role as the Director of BARC, Raja Ramanna was given much credit. Some opine that it was the 'extra familiar'[2] relationship Ramanna shared with Indira Gandhi that enabled

2 Interview by G Perkovich with former high-ranking AEC and defence officials, January 1996 and September 1997. *India's Nuclear Bomb*, p. 521.

the nuclear establishment to convince the Prime Minister for the test despite 'poor performance and budgetary travails' of the DAE during those times (Perkovich, 1999:521). He gave essential inputs in all the decisive meetings that took place before the test. Indira Gandhi 'simply trusted him and went with his advice' (Ibid: 176). By then, Ramanna had made site visits while preparations were on at the Thar desert. Writing about his experience, he states:

> An unexpected bonus as we readied the site for the experiment was our discovery of the great beauty of the desert and its denizens. The Thar bubbled with life. Magnificent peacocks and small spotted deer followed us frequently...The jawans served us wholesome and healthy food...Though it was summer we got used to the intense summer heat during the day and looked forward to the nights which were pleasant.
>
> (1991:91)

With everything in place, the end result of all the efforts by the strategic enclave is a story well known today – Pokhran-1, India's first underground nuclear test, where three fission devices were successfully detonated. In Ramanna's words 'right in front of [us], the whole earth rose up as though Lord Hanuman had lifted it' (Ibid.).

Raja Ramanna had reached a high point in his career as a scientist by early 1970s. During those times and onwards, India had to resist the pressure from the West regarding safeguards. Ramanna did so as a part of the regular Indian delegation at the Geneva Conference during the late 1960s and early 1970s.

His contributions to India's nuclear programme were talked about. The Government of India honoured him with the Padma Bhushan Award in 1973. Despite being an adamant advocate of a nuclear test, by no means was he turning into a 'pro-bomb nuclear hawk' because it was only to its peaceful use that Ramanna was attracted. He would admit later how a nuclear weapon was a devil with which everyone had to live, and ever enthusiastically bring out the peaceful nature of the atom for the twenty-first century.

To what extent he wanted a nuclear device is interesting to analyse. As late as early 1974, Ramanna thought the reason a test was inevitable was because 'it was now impossible to postpone the date given the expenses, time and the critical stage the experiment had reached' (Ramanna, 1991:89). Yet there were other reasons why he thought it was irrevocable. He had no doubt about the benefits that would follow a PNE. That India was capable of having its own successful PNE was irresistible. He tremendously believed that it was necessary for India to prove its credibility and enhance its status vis-à-vis the colonial pressure of the non-proliferation regime. This was indeed his motivation for a test then and later. His advocacy for nuclear tests for India thus has to be viewed under a scientific and nationalistic light.

As years passed, Ramanna grew more convinced about the usefulness of nuclear energy and discriminatory nature of NPT. He vehemently opposed the NPT, and thought it was unfortunate that attempts to prevent proliferation were being made on the basis of embargoes, punishments and by use of force. Equating it to a form of 'punitive sanction' he said:

> [P]unishment even in medieval times referred to a crime but now it seems punishment can come even before the crime...If non-proliferation has to succeed it can only be in a voluntary manner and joint action is essential.
>
> (Chettyar Memorial Lecture, 1992)[3]

After the Pokhran test, along with R Chidambaram, he prepared a paper titled 'Some Studies on India's Peaceful Nuclear Explosion Experiment' for the IAEA meet in January 1975. Stating the yield of the Pokhran explosion to be twelve kilo-ton, they outlined the aim of the Indian nuclear programme as having civilian applications alone. After Pokhran, though many questioned their claims, and ambiguity reflected in the answers, Ramanna managed to clear the case for India at the IAEA to a great extent.

Dr Raja Ramanna's contribution to making Operation Smiling Buddha possible in 1974 led many to attribute immense credit to him for the bomb. Given the scientist's inclination away from a 'bomb' in the sense of a military nuclear weapon, this privileging is questionable whatsoever. When asked about Raja Ramanna, Dr G Perkovich fondly narrates about how the man, full of mischief, worried about the evils of the Indian caste system and the dangers of giving control of the nuclear programme to the army (Interaction with the author during a

3 *U.S.I Journal,* (1992), 'Security, Deterrence and the Future', Text of Chidambaram Chettyar Memorial Lecture delivered on 8 July 1992.

seminar in JNU, New Delhi in November 2015). Ramanna himself warned about the dangers of the nuclear weapon; he thought of it as most dangerous, as terrorists could someday access it and cause devastation. However, he had 'no doubt that atomic energy in the next century will play a bigger role than ever before' (Interview with Rajiv Mehrotra).

While he was the Director at BARC, Dr Ramanna (and few others) proposed a stochastic theory of fission with the explanation of 'random walk' in a paper titled 'The Angular Correlation of Prompt Neutrons and Fission Fragments in Thermal Neutron Fission of U235' (Ramanna, 1991:80-81). But, besides such work, Ramanna also encouraged scientific research in other fields like studying solid waste matter, material and metallurgical science, use of radiation for food preservation, nuclear energy for power generation and many other experimental activities. For all his contributions to the scientific progress of the country, Dr Ramanna received the Padma Vibhushan Award in 1975.

The nuclear reactor at Trombay, CIRUS, had begun to age. To replace this, AEC had begun construction of the R-5 100 MW reactor (Dhruva) by 1977. The same year also witnessed removal of Ramanna's benefactor Indira Gandhi from power. Morarji Desai (more Gandhian than Indira Gandhi!) stood for all that was anti-nuclear. He even dismissed the need for PNE as 'politically motivated' (Perkovich,1999:531). Certainly, this caused some ripples within the nuclear establishments. Dr Ramanna did look up to the new Prime Minster for his views on NPT and his refusal to sign it, but his equations with neither Desai nor Sethna were perfect – the former wished to cap the nuclear activities under his leadership while the latter had an envious relation with Ramanna over Pokhran's credits (Ibid., p. 214). Dr Raja Ramanna was found low in spirits and health during this phase. He awaited a change in environment and soon it came in the form of a transfer to New Delhi to shoulder new responsibilities as Secretary of Defence Research and also Director General, DRDO.

BARC under Dr Ramanna (during his first innings as Director from 1972-78) can be summed up in his own words:

> The period was one which saw several successes, but it was also a time of trials and tribulations. It was during this period that BARC reached its acme both nationally and internationally. It had gained enough confidence not only in the servicing of power reactors but was also able to design new types of research reactors of larger size.
>
> (1991:96)

Ramanna from 1978-1981

Prior to Ramanna's transfer, there existed only three departments within the Ministry of Defence. Department of Defence Research and Development was created in 1978 and Ramanna was placed as its first Secretary, in turn becoming the scientific advisor to the Prime Minister, the same person who had to pluck Ramanna out of BARC to cool the spirit of the pro-explosive leaders in Trombay (Perkovich,1999:214). If this is taken to be true, then the kind of authority and aura Ramanna had at BARC can only be estimated. In fact, Homi

Sethna's attempt to find a permanent replacement for Ramanna was vetoed by most in BARC (Ibid: 228).

Dr Ramanna received an 'obsequious welcome' but soon found strings trying to hinder his efforts. He was not included in the Defence Planning Board; and it was communicated that his department though independent was to be seen as a sub-unit of the defence production department. He made suggestions to improve research in the defence department which was seen as a sham without any coordination and 'total dependence on outside know-how'. Ramanna felt the place reeked of political ploy when some people asked him to give up the post of Secretary; he knew that in Delhi 'if one was not a Secretary to the Government, then one was reduced to being a non-entity' (Ramanna, 1991:97-98). He also witnessed the bureaucratic style of functioning in Delhi but nevertheless did his best. At the DRDO, he cancelled as many as five hundred inefficient projects to increase concentration on a few major ones in order to achieve self-sufficiency (Ibid: 103). Here too troubles came following, but, with help of advisors and well-wishers, Raja Ramanna tackled his job meticulously and with great calm.

Dr Ramanna had already spent much time with science and had missed out on his music and travel to a great extent. Delhi did offer him much scope to travel if not music. An interesting anecdote is his visit to Iraq as a personal guest of then President, Saddam Hussein, who had taken an interest in the career of this chief nuclear physicist of 'India's bomb'. During this 1978 visit, Saddam Hussein insisted Dr Raja Ramanna stay back in Iraq and help build its nuclear programme. The dictator in fact uttered, 'I will pay you whatever you want. I expect you to honour this offer'. A worried man, Ramanna hurriedly flew back and never went to Iraq again! (*The Times of India*, 25 September 2004)

Back to Trombay

In January 1981, Dr Raja Ramanna was recalled to BARC to resume his post as Director and with additional responsibility of Secretary to the Government's Department of Atomic Energy as well as Scientific Advisor to the Defence Ministry in New Delhi (succeeded by VS Arunachalam in 1982). His increased stature also corresponded with Indira Gandhi's return to power the previous year. Since the establishment had been neglected under the Desai government, Ramanna worked to invigorate it with new life. He could now work without much interference and chalked out two goals for the near future – completion of Dhruva reactor and getting the Fast Breeder Test Reactor (FBTR) at Kalpakkam to achieve criticality. Ramanna's second innings at BARC was a full three-year term and was as crucial as his first.

BARC and DRDO were aware of the need to have more PNE experiments if the country was to produce deliverable reliable nuclear weapons for counter strikes. Work to better the technology had been going on since post-Pokhran. BARC had been involved in increasing the efficiency of neutron initiator, a project on which Ramanna took great interest. More 'pro-bomb' scientists were making their own efforts too and by early 1980s felt the need to test their work. An avenue was sought to voice their interest. Dr Ramanna and Dr Arunachalam, sometime in early 1983, managed to present their case with a request for another nuclear experiment before the Prime Minister and spoke about its technical necessity. Indeed, a second nuclear test was approved by Indira Gandhi in 1983, which however was revoked within twenty-four hours. Dr Ramanna and others were never told why.

As previously mentioned, for Raja Ramanna, the hope to use nuclear energy for the progress of the country and its self-reliance stood special. It was he who mooted the idea of having 10,000 MW of nuclear power capacity by the turn of the century, an ambitious yet novel vision. The Government endorsed this idea as a fifteen-year programme for nuclear power in India. Many doubted the attainability of this aim but the man was vociferous in his propagation of the plan. In an interview in March 1984, he said;

> Many people feel that just because we are at 1,100 MW, extrapolating to 10,000 MW means the same kind of effort that went into reaching the 1,000 MW level. As a matter of fact, if you take the decision to put up a large number of reactors, fabricating of components becomes much easier: It is more a question of repeatability, not of designing new things.
>
> (Strategic Digest, 1984:320)

Further, Ramanna opposed the government's plan to meet targets by importing reactors. Though he was assured it was only to add to indigenous efforts, Ramanna was of the view that Indian industries would be able to support the plan without the need for imports. He was positive about it, and looked at import as the last resort in a situation where time ran out (Ibid.). His adamant adherence to self-reliance made him clash with Rajiv Gandhi by 1985, a clamour very well felt by people around. In his autobiography, Ramanna writes about being so concerned by the Government's manoeuvrings that he hoped for retirement (Ramanna,1991:115)! Ramanna meanwhile conducted studies on how India could go ahead with such a programme without undue scrutiny of safeguards. With a workable blueprint in hand, he now put his force behind accelerating the work on Dhruva and getting the FBTR ready.

By 1981, four years after its inception, Dhruva reactor with a capacity of 100 MW was lagging behind schedule. Dr Ramanna's return gave it a new lease of life. Placed next to CIRUS, it achieved criticality in 1983 and gave the country another reactor, which produced most isotopes for it. Ramanna heralded the reactor as 'probably the best reactor built in India' (Interview with Rajiv Mehrotra). It was a reactor free from safeguards but emphasis on its civilian application was made. The same year the unsafeguarded nuclear plant in Madras was commissioned.

In 1983, Ramanna was named Chairman of AEC. In this capacity, Dr Ramanna, with the help of N Srinivasan, pushed the FBTR to completion. It went critical on 18 October 1985 making India the seventh country in the world to master the fast breeder technology. Ramanna, to whom its completion is attributed, remarked, 'the satisfactory functioning of the

FBTR will go down as a landmark in the history of nuclear India' (Ramanna, 1991:112).

Sixty years old and smeared with wisdom, by 1985, Raja Ramanna's career had already become an inspiration for many scientists in India. His propagation of nuclear energy for generating power (as best for India), making use of the atom for various civil purposes, and such other advanced ideas comprised bulk of his writings and speeches from the mid-1980s. He clearly brought out how isotopes could be used in agriculture by production of mutants or irradiation used for preservation of food, its use in treatment of cancer or for sterilization purposes in the health sector, use of radiation for detecting defects in mechanical parts in the engineering sector and so on. He was awarded the Meghnad Saha Medal of the Indian National Science Academy in 1984, the Om Prakash Bhasin Award in 1985 and the RD Birla Memorial Award in 1986. Thus, with all the honour and glory of his contribution as a physicist and nuclear scientist, Dr Raja Ramanna retired as AEC Chairman in 1987. His teachings, efforts and visions have indeed been a motivation for many aspiring scientists.

Other involvements

If science was his vocation, music was no pastime for Dr Raja Ramanna. He was a trained pianist and spared no opportunity to enhance his skill. Whenever he travelled (especially to Europe), more often than not, he somehow managed to find time to appreciate and enjoy the music of the place. He himself gave quite a few performances in India and received great applause for it. With age, Ramanna began to associate music with meditation, a form of yoga (Interview with Rajiv Mehrotra). He authored a book on music titled *The Structure of Music in Raga and Western Systems,* published in 1993.

Raja Ramanna further took interest in Sanskrit and leaned towards philosophy during his later years. Advaita and its teachings reflected in his conversations. He explained how Buddhism was akin to science as it looked for causation and did not believe in soul theory (Ibid.). Science for Ramanna further stood as an instrument of social and economic change, but in his own words he also stated:

> I would like to look upon science, first of all as a philosophy; it is a part of philosophy. Many people tell me that science is separate and philosophy is separate. Perhaps religion is separate, but philosophy is not. In the West, they call it 'scientific philosophy'.
>
> (Ramanna, 1992: 83)

Ramanna also made important contributions to enhance scientific research in India. As early as 1969, he visited USA and other places with the aim of setting up the Variable Energy Cyclotron Centre (VECC) at Calcutta. Today the VECC is a frontier research and development institute which carries out work in accelerator science and technology, nuclear material and computer sciences and other relevant areas. As President of the Indian National Science Academy in 1977-78, he brought about significant changes in the recruitment processes of the institution as well as worked towards improving the quality of publications which was considered 'sub-standard'. His efforts were however resisted by the older members of the institution and this gradually made Ramanna lose interest in it (Ramanna,1991:119-20).

In 1984, he worked towards the establishment of the Centre for Advanced Technology at Indore. Recognizing his role in it, this centre was renamed as Raja Ramanna Centre for Advanced Technology. It is a part of DAE and conducts research on non-nuclear areas of lasers, particle accelerators and other such technologies. On the initiative of JRD Tata, Ramanna worked out the idea of the National Institute of Advanced Studies (NIAS) in 1987 and became its founding Director. Its aims were:

> One, to include in the curriculum some selected aspects of humanities. Second, it would provide a comprehensive course for senior executives and government officials, to broaden their existing attitude.
>
> (Ibid: 162)

Ramanna, throughout his career, became associated with many institutions. He was the Chairman, Governing Council in the Indian Institute of Science, Bangalore; Council of Management, Jawaharlal Nehru Centre for Advanced Scientific Research, Bangalore; Chairman, Board of Governors, Indian Institute of Technology, Bombay (1972-78) and many other such establishments. He further became President of the 30th General Conference of the International Atomic Energy Agency at Vienna in 1986 and Vice President of Indian Academy of

Sciences. Whatever role he played, Dr Ramanna did so with ease and rarely failed in his efforts.

Later life and end of his pilgrimage

One morning, in December 1989, while going through his routine work at NIAS, Dr Ramanna received a call from then Prime Minister VP Singh's office. On being summoned to Delhi, he wondered what was in store this time. Ten days after his meeting with the Prime Minister, on 21 January 1990, Dr Raja Ramanna was sworn in as the Minister of State for Defence. His post-retirement life took a new turn. As a minister and as a politician too, Ramanna earned respect for himself.

India's nuclear programme gained some pace in the 1990s. Besides other matters, Dr Ramanna's views were sought on most issues dealing with the nuclear programme. Now a politician, and having aged with time, he had come to the conclusion that there was no need to have any thermonuclear test for the sake of increasing deterrence. The scientist in him did support experiments however. In 1985, when Rajiv Gandhi had tasked him (and others) to consider India's defence planning needs, the group had recommended that a minimal deterrent force would be sufficient (Perkovich,1999: 274). Reiterating this view in 1996, he said that the 'plutonium bomb is sufficient enough to act as a deterrent and we have the capability' (Ibid: 384). There was thus no need for any thermonuclear test. Instead India should concentrate on enhancing its nuclear energy production. He wrote:

> A small number of nuclear weapons are sufficient for mutual annihilation of countries and this itself is sufficient deterrence to cast doubts on the outcome of a nuclear military adventure. The possession of nuclear weapons also gives confidence to the conventional forces, in that it confines the possible worst outcome of a war, to only a stalemate.

Ramanna was confident of India's progress towards self-sufficiency and even suggested that its power reactors could be placed under IAEA safeguards; thus the country could gain tremendously for being a responsible power. By stating that India had enough plutonium, he indicated that he was mostly content about the nuclear path the country was taking.

In 1996, Ramanna received the Asutosh Mookerjee Gold Medal. He was further nominated to serve as a member of Rajya Sabha. During his term from 1997-2003, he participated fruitfully in various debates and discussions.

In September 2004, while visiting his son in Mumbai, Dr Ramanna complained of feeling unwell and uneasy. He was soon admitted to the Bombay Hospital. Two days later, on 24 September, he passed away due to a cardiac arrest. The beautiful and inspiring 'pilgrimage' of Dr Raja Ramanna came to an end. The scientific community mourned the loss of the 'Father of India's Nuclear Programme', an adage attributed to him for his ocean of contributions which made India stand tall in the world of nuclear science.

Conclusion

Dr Raja Ramanna's journey of life through the avenues of science, music and philosophy, and his experiences of varied places is a story that draws attention. Whether it is his contribution to the PNE of 1974, his vision of the fifteen-year nuclear power programme in 1985, his obstinate views about self-sufficiency for India or propagation of the peaceful uses of atom, it has all been of great relevance for modern India. Ramanna's love for music and his piano added a vibrant and artistic side to his dynamic personality. During his later years, he grew inquisitive about the philosophy of 'being', and Advaita became his area of interest.

It is unfortunate that today when a question is asked about people who contributed to India's nuclear and scientific progress, only the names of Homi J Bhabha, Dr APJ Abdul Kalam and Vikram Sarabhai come to mind instantly. Only after the initial round and some thought, do few people recall the name of Dr Raja Ramanna, despite his truly great achievements. It is indeed wished that more people than those just from the relevant fields recognize the immense contributions that Dr Ramanna made to India's nuclear programme and otherwise. It is with such hope that the story of Raja Ramanna needs to be told.

REFERENCES

- Ramanna, R. *Journal of the United Service Institution of India* 'Security, Deterrence and the Future', Text of Chidambaram Chettyar Memorial Lecture delivered on 8 July 1992.
- Ramanna, R. 'Security, Deterrence and the Future', *Economic and Political Weekly*, Vol. 27, No. 37 (Sep. 12, 1992), p. 1975-1978.
- Ramanna, R. *Journal of Indian Ocean Studies* 'Technology Transfer and Non-Proliferation', NIAS, Volume 1 Number 1, November 1993.
- Ramanna, R. *Years of Pilgrimage: An Autobiography*, Viking, 1991.

1. Raja Ramanna with PM Indira Gandhi 2. Raja Ramanna

APJ ABDUL KALAM
The Missile Man of India

Rajaram Nagappa

MUCH HAS BEEN written about APJ Abdul Kalam in the last few years since his passing. Scientists, technologists, bureaucrats, politicians, teachers and his most ardent fans – students – have vied with each other in talking about his sterling qualities, and how his words and deeds have continued to inspire them. Rich tributes have been paid to him by national and international leaders, by scores of people whom he seems to have touched in special and humane ways, and by a galaxy of students, for many of whom Abdul Kalam was a role model. The National Space Society in Washington DC, which had honoured Kalam with the prestigious Wernher von Braun Memorial Award in 2013, in its obituary note had the headline 'One of our Leading Lights has joined the Stars'. Dr Kalam was already a star in his lifetime and his achievements enveloped a very wide and diverse canvas. The width and depth of his interests spanning science, technology, arts, education, music, his passion for teaching and communication especially with the student community, his ever-present optimism, his nobility and the inspirational leadership he provided has been widely acclaimed and written about.

There is not much that I can add to the already rich, spontaneous and well-written lore on Abdul Kalam. What I can capture though is his foresight and passion for technology, his technology management skills, his confidence in himself and in his team, and his passion to explore realms outside the scope of technology. I will also touch upon his passion for learning from one and all, his total lack of airs and his accessibility, even when he held high offices.

At a personal level, though my association with Dr Kalam started when I joined the Space Science and Technology Centre (the predecessor organization to the Vikram Sarabhai Space Centre) at Thiruvananthapuram, in January 1967, my interaction with him continued even after he left ISRO in 1982 to lead the Defence Research and Development Organisation to new heights and in subsequent years also. Through this experience of knowing him as a senior, a colleague, an excellent scientist and a fine human being, I would like to pen down the story of Kalam, the scientist President of nuclear and modern India.

Childhood and adolescence

Little Kalam was born on 15 October 1931 in Rameswaram, Tamil Nadu, when India was sixteen years away from getting independence. He was born in a normal middle-class family with his father having a good social circle and a boat building business. His father Jainulabdeen Marakayar had a significant influence on Kalam. The youngest of five siblings, Kalam was always inquisitive about his surroundings and often asked his father importance of things like prayer, unity and nature. Ahmed Jallaluddin, Kalam's brother in-law called him *Azad*; it was with him that Kalam often discussed spiritual matters, scientific discoveries, literature and medical science.

He was just eight years old when World War II started. That was the time when Kalam had his first job as a tamarind seed collector for which he earned one *anna* a day; he later worked as a newspaper boy during the state of emergency when India joined allied forces during World War II. Kalam, in his autobiography *Wings of Fire*, talks about 'three wise men'. Once Kalam went to school wearing his prayer cap, following which the teacher made him sit away from his three close friends, Ramanadha Sastry, Aravindan, and Sivaprakasan, who were from orthodox Hindu families. After this incident, they went to their respective fathers and told them about the teacher's behaviour. When Kalam's father heard about it, he decided to visit the school along with a Brahmin friend and a Christian priest, and requested that kids should not be discriminated based on caste or religion. This story of the 'three wise men' Hindu, Muslim and Christian had a huge impact on Kalam and he understood the uniqueness of India. His science teacher Sivasubramania Iyer once said, 'Kalam, I want you to develop so that you are on par with the highly educated people of the big cities (Kalam, 1999).'

Kalam was always an average student as far as grades were

concerned, but he was sharp and hardworking. He had a desire to learn rather than parrot the syllabus. He worked hard to get educated in the best schools of that time like Schwartz Higher Secondary School, St Joseph's College, and University of Madras from where he graduated in physics in 1954; later, in 1955, he studied Aerospace Engineering in Madras Institute of Technology.

My personal sojourn with an ardent scientist

I joined the Space Science and Technology Centre (SSTC) in January 1967. I had obtained my diploma in aeronautical (often shortened to 'aero') science from the Madras Institute of Technology (MIT). The Thumba campus boasted of a few names from MIT on its rolls. Among them were MC Mathur and R Vivekanandan from aero, B Ramakrishna Rao and R Aravamudan from electronics, and the popular MITian, Abdul Kalam. Kalam was already known to the aeronautics community. Thumba campus was developed on a church land in a hamlet called Thumba in the coastal city of Thiruvananthapuram in Kerala. It was converted into a Space Centre, which was founded by two visionary scientists Home Bhabha and Vikram Sarabhai. Kalam had developed and flown India's first hovercraft while working as a Scientific Officer in the Aeronautical Development Establishment, Bangalore, prior to joining Thumba Equatorial Rocket Launching Station (TERLS). My first interaction with Kalam was on the afternoon when I walked into his office across the Church Building to meet with him and introduce myself. He was talking to one of his engineers, V Sudhakar, as they were sorting out pyrotechnic pellets. There were a couple of other engineers working on hardware. The atmosphere was friendly, informal and a bit chaotic and Kalam seemed to be in the middle of everything.

Kalam was among the initial batch of engineers picked up by Dr Vikram Sarabhai in the early 1960's for India's fledgling rocket programme operating from the newly started Thumba Equatorial Rocket Launching Station. On his return to India after an initial period of training at NASA's Wallops Flight Facility, Kalam was put in charge of the Rocket Engineering Division. He was responsible for payload integration and developing various mechanisms for deploying the payload. Based on a need for radio signal transparency during 1966-67 for some payload, Kalam and his team started developing nose cones made of Glass Fibre Reinforced Polymer (GFRP). This was the first time when glass fibre was being put to use in a hi-tech area in India. Right from his early days in rocketry, Kalam strove to push technological innovation and new ideas. He wanted to ensure that technology advancement took place with the constant development of rockets and rocket subsystems. Also, hi-tech space technology developments are associated with risk-taking and are a highly visible activity. Kalam ensured that latest technology development found place in every new work that was undertaken by him. This trait was very much on display when the Indian Air Force requested ISRO to develop a rocket motor (a rocket power plant using solid propellants is referred to as motor) for Rocket-Assisted Take-Off (RATO) of Su-7 and HF-24 Marut aircrafts. The idea was to reduce the take-off distance of the aircraft as well as cut costs by import substitution. Kalam was named to head the project. He took on the job with utmost confidence in the capability of his colleagues to meet the requirement. Considering the fact that the whole of indigenous rocket technology was at a stage of infancy, methods, procedures and quality issues were just evolving, facilities were yet to be established, industrial support in Trivandrum was wanting, it was indeed a bold move to take on the challenge of developing the RATO system keeping in mind it was mounted on to a costly asset of the IAF and there was a pilot on board the aircraft. This was an early manifestation of Kalam's approach to technology development – technology development without being risk-averse. Development of a system like RATO involved risks – risks of not meeting the specifications, risks of time and cost overruns, risks of managing the integration and interfaces with the aircraft platform and other related issues. Kalam's approach was that risks could be weighed, analysed and appropriately addressed without compromising on safety of the test article and the platform. In adopting this approach, Kalam was not only having self-confidence in meeting the development objectives, but he also had the confidence and trust of the team he was leading in meeting the goals. In fact, this was a trait that could be repeatedly seen in any new development activity that Kalam undertook.

At the time of RATO development, we had indigenously developed and tested the RH-75 rocket. The rocket was 75 mm in diameter with no defined performance requirement except that it should demonstrate successful flight capability. The development was taken up essentially as a technology learning project and it provided the essential inputs related to design, analysis and performance estimation. More importantly, it provided inputs in sealing the joints. In addition, we had the Centaure sounding rocket technology acquired from Sud Aviation, France. With this fledgling information, the design of the RATO motor was taken up. The design and realization of the RATO motor involved the following challenges:

- The motor was not of uniform diameter. The forward end was required to be of a lower diameter to match with the support mount on the aircraft.
- The exhaust nozzle had to be canted by 24° so that the motor thrust axis would be aligned with the centre of gravity of the aircraft.
- The nozzle exhaust should be directed away from the aircraft.
- A safety feature had to be included to ensure relief in case of pressure excursion from the nominal.

As a technology improvement measure, it was decided to use glass fibre reinforced plastic for the cylindrical part of the dome and metal for the forward end of the closure, which had a smaller diameter.

Doggedness to solve an issue rather than bypass it or give up was an ISRO characteristic, and Kalam was an ardent advocate and follower of this approach. During the qualification tests of RATO, a major issue arose when the aft-end dome failed. The test was carried out in the afternoon and we had to wait for some time to approach the test stand to examine the debris. Kalam suggested we meet immediately to take stock of the situation and plan the further course of action. Our discussion lasted till the early hours of the morning, but we managed to get a grip on the problem and the remedial actions we needed to put into place. The corrective actions started in full swing the next morning, about six hours after our discussion and decision. Overall quality control and inspection standards were also tightened and radiographic inspection of the propellant grain was added. From reliability considerations, we had to repeat a few tests, before clearance could be given for mating with the aircraft. We successfully demonstrated the functioning of the motor on ground and in aircraft taxiing trials.

Kalam was a pusher of new technology initiatives and never lost an opportunity to include a technology element that would enhance the performance of the system. Very early in the rocket systems development, he saw the advantage of using reinforced plastics as structural elements to reap the benefit of reduced inert mass.

Satellite launch vehicle development studies in ISRO were undertaken as early as 1968 and the configuration of SLV-3 as a four-stage rocket was finalised after studying alternative configurations. When Vikram Sarabhai initiated the development of SLV-3, he entrusted the responsibility of development of each of the four stages of the vehicle to different specialists. The fourth stage as conceived at that time was 560 mm in diameter, and Kalam was given the responsibility for developing the stage.

Kalam decided that the motor case material will be glass fibre reinforced polymer (GFRP). This made sense as any inert mass saved in the top stage would directly translate into additional payload and a good GFRP design could be twenty to thirty percent lower than a metallic case.

Just like the development of RATO motor, the development of the S-4 stage motor would have gone through with phases of technology push and risk assessment, but what stands out is the alacrity with which Kalam could seize the opportunity and momentum to push development into higher scales. At about this time, France was developing a launch vehicle named Diamant BP4, which employed a solid propellant top stage of 657 mm diameter – slightly bigger than the SLV S4 motor. The French team was in Trivandrum making a presentation of their launch vehicle. Kalam saw the opportunity for fielding the SLV S4 motor as a candidate for use in Diamant BP4, and immediately offered to modify the design of SLV-3 fourth stage motor to meet the French requirements. Overnight, back-of-the-envelope calculations and hurried sketches of the modified motor were made and presented to the French. Unfortunately, the French cancelled the development of their vehicle and an opportunity for marketing the S4 motor was missed. However, a few years later, the S4 motor did fly on the Ariane-4 rocket as the Apogee motor of India's experimental communication satellite, Ariane Passenger Payload Experiment (APPLE).

Kalam was a voracious reader. Ayn Rand was his favourite author, and he would recommend her novel *Anthem* to all of us. His non-fiction reading centred around management books;

Kalam is one of the few people I know who not only read but also applied relevant elements to manage his projects and interpersonal relations. Kalam used to be extremely focused on achieving the end result and dexterously handled technology, personnel and management issues. The Satellite Launch Vehicle, SLV-3, was a development project where each of the forty-two major subsystems had to be *de novo* designed, developed and realized by different teams. The project also involved creation of facilities and infrastructure. The task of managing diverse teams, spread over ISRO centres, was challenging to say the least. Kalam managed the tasks of bridging priority, resource balancing, funding and the normal inter-personal issues with dexterity. His judgment was accepted as it was fair, had no element of favouritism and was meant only to advance the cause of the project.

The Polar Satellite Launch Vehicle (PSLV) has proved to be a reliable launch vehicle notching up over forty flights. At the time of finalization of the PSLV configuration, there were other contending configurations and their votaries. It was Dr Kalam, looking after the ISRO launch vehicle programme office at that time, who studied the different configurations and recommended the current vehicle comprising two solid propellant stages and two liquid propellant stages for the main propulsion. The soundness and the wisdom of the choice has stood the test of time, and today PSLV and its variants have proved to be the workhorse launch vehicle of ISRO.

Another attribute that Kalam possessed was not being afraid of failure. In a development project, all problems and failure modes cannot be completely anticipated. Kalam, more than anyone else, was well aware of this and took failures in his stride. He strongly felt that failures provide the stepping stones to success – except one must carefully analyse the failures, identify the causes, incorporate remedial measures after confirming their efficacy. This involved getting to the root of the problem and rectifying it; then there was no room for recurrence of the problem.

The Defence Research and Development Organisation (DRDO) had initiated work on development of missiles in 1956 by the Special Weapons Development Team (SWDT). The SWDT transformed into Defence Research and Development Laboratory (DRDL). It was responsible for the development of a wire-guided anti-tank missile, indigenisation of the Russian SA-75 missile, code named 'Devil', and other missile systems. In 1974, a Review Committee under the Chairmanship of Dr Brahm Prakash, Director of Vikram Sarabhai Space Centre, made detailed recommendations for DRDO in respect of systems analysis, system specification and technology development. The committee recommendations also included facility requirements towards hardware realization, testing, computer centre and quality assurance. Project management needs were also spelt out in the recommendations. Abdul Kalam was a member of the Committee.

At about this time, a special committee under the chairmanship of Professor MGK Menon made comprehensive recommendations related to requirements for all classes of missiles to be developed, their performance characteristics, the technologies to be developed and/or acquired from abroad and the production needs.

National security was high on Kalam's mind, and he rightly believed that true deterrence comes when one has not only strategic weapon capability, but also the means to deliver it on targets deep within enemy territory. In the early eighties, Kalam moved from ISRO to DRDO and his stamp of an original approach to missile development was immediately visible. The MGK Menon Committee report provided the backdrop for starting the Integrated Missile Development Programme (IGMDP). In ISRO, the result of the technology development was to primarily meet the needs specified by ISRO; compromises were not difficult in case of system performance shortfalls. The products of DRDO were however tuned to meet the needs of defence forces, and a set of specifications agreed to at the start of development were sacrosanct. Kalam's message to the user agencies was that DRDO could not meet their today's needs but could develop systems needed by them at a later date. He therefore demanded that they forecast their requirements regarding weapons and weapon performance goals five years into the future. The developments under IGMDP were based on this precept and comprised the following, a) anti-tank missile *Nag*, b) surface-air-air missile *Akash*, c) a quick reaction surface to air missile/ship launched anti-sea skimmer missile *Trishul*, d) a surface-to-surface battlefield missile *Prithvi* and e) a re-entry technology demonstrator *Agni*.

In the management of IGMDP, Kalam brought in decentralisation in decision making and introduced concurrent engineering. The designers, users, fabricators and industry were brought together to ensure technical and timely coordination. In this process, the designer was made to appreciate the practical limitations posed by process and fabrication of equipment/ processes and keep appropriate design allowances. Similar interaction followed in respect of material availability and downstream conversion processes, thus integration as well as handling and transportation requirements could be built into the design. The process of concurrent engineering brought all the stakeholder's requirements to a common meeting ground, leading to early recognition of issues with resultant solutions and time saving.

Kalam also brought technology management skills honed in the realization of SLV-3 to bear on the IGMDP projects. Documentation and introduction of design reviews ensured cross-fertilization of ideas among competing groups. Though liquid propulsion systems are not the prime options for ballistic missiles, keeping in mind the progress achieved in Project Devil, Kalam encouraged its adoption for the development of Prithvi missile. The Agni Technology Demonstrator (Agni-TD), a two-stage vehicle could be developed after the demonstration phase ballistic missiles of 1000 km range, and hence early realisation of the system was important. Kalam, therefore, turned to available systems and configured Agni-TD with the first stage fashioned after SLV-3 booster stage, and second stage based on Prithvi. This decision resulted in substantial time saving. Kalam made sure that the Indian industry was involved to a significant extent in production of missile components and subsystems. Under his stewardship, missile development took major strides, and the reach of Indian missiles is long enough to discourage any misadventure by adversaries and add to the deterrence capability. This is indeed the unsaid contribution of Kalam to India's nuclear delivery system, beyond the various direct involvements of his in both Operation Smiling Buddha in 1974 and Operation Shakti in 1998.

Today India is one of the few countries possessing a supersonic cruise missile. The speed and low altitude flight profile of BrahMos cruise missile makes it a major and hard to defend against weapon system. Kalam recognised the effectiveness of this system and entered into a very successful joint venture agreement with Russia. The system has been made multi-platform compatible and is a major asset to our defence forces.

From all accounts, the Indian missile programme is completely indigenous and a commendable achievement. The achievement becomes all the more important as the developments have been carried out under severe adverse conditions of sanctions and technology denial regimes like Missile Technology Control Regime (MTCR) imposed on DRDO and other entities in the country. Kalam believed strength respects strength as evident from his response to a query from a reporter in 1994 on the destructive capability of missiles. He said, 'by building such arsenal, I actually ensure peace for my country. Now no nation dare attack us. These are truly weapons of peace'. Kalam, the Missile Man of India, truly succeeded in putting India in a position of undeniable strength with his weapons of peace.

Kalam's other pursuits

While I have described Kalam's contributions to the development of missiles in DRDO, his contributions to other spheres of DRDO were equally significant. He steered the development of the Light Combat Aircraft – a multi system man-rated supersonic combat aircraft – right up to the roll out of the TD-1 in November 1995. Embargos imposed by the US after the 1998 Pokhran tests, as well as technical complexities, delayed the project and the maiden flight of the aircraft could only take place in January 2001, after he had demitted office of Scientific Advisor to *Raksha Mantri* (Defence Minister). In this context, one of his unfulfilled dreams related to the development of a 100-seater civil regional transport aircraft. He felt a country like India should be in a position to build and fly its own civil transport. When he was President of the Aeronautical Society of India, he tried his best to reorganise the aeronautical setup in India. The report of a committee he had set up for the purpose was widely discussed in the Aeronautical Society and submitted to the Government of India, but unfortunately saw no progress.

In some aspects of technology advancement, Kalam was far ahead of his time. He saw major promise in the development of reusable hypersonic airbreathing rocket systems as an economic way to access space. His engineers studied in detail means to achieve orbit using a single-stage rocket (SSTO) using combined cycle propulsion along with a concept of collecting air during the atmospheric phase of flight and liquefying it on-board for use as oxidiser in the exo-atmospheric phase of flight. Though the concept held a lot of promise, it was ahead of its time – a number of material and technology related issues had to be studied and the cost factor, including the cost of test facilities, was quite high. To add to it, launch service providers had made huge investments in expendable launch vehicles (ELVs) and were not keen on investing in new technologies till they had recovered the costs they had spent on ELVs.

It was during his chairmanship of the Technology Information Forecasting and Assessment Council (TIFAC) that important time-bound missions like the sugar mission, oil mission, fly ash mission and composites mission were pursued. He advocated harnessing the power of technology for societal good. Among his priorities were agriculture, food processing, providing urban facilities in rural areas, infrastructure, information and communication, quality and affordable education. Educating children and inculcating values in them was a passion and he made it a point to reach out to children in all parts of India.

Kalam held his teachers in high esteem. He was in constant touch with Professor KAV Pandalai, who had taught him aeronautics at the Madras Institute of Technology (MIT). In one meeting where both Professor Pandalai and his music teacher were present, he started the meeting offering his *pranams* to both his gurus. He was instrumental in instituting a Chair – Professor KAV Pandalai Memorial AR&DB Chair – in Professor Pandalai's memory at the Faculty of Aeronautics in MIT in 1999. The author had the privilege of being the first to serve as this Chair Professor.

Kalam's role models were Dr Brahm Prakash, who was the Director of Vikram Sarabhai Space Centre during the SLV-3 development days, and Professor Satish Dhawan, who was the then Chairman of ISRO. He not only held them in high esteem, but also emulated them in his pursuit for perfection.

Honours and awards sat very lightly on him. He was never to flaunt any of his accomplishments. He would humbly say the recognition bestowed upon him was due to the hard work of his colleagues, and the honours belonged to all of them. India's top civilian award – Bharat Ratna – was bestowed on him in 1997. The award helped him travel the length and breadth of India and interact with students all over the country. He had a way with students, and students found great inspiration in his work and deeds. This became evident over time when many freshers I used to meet at MIT, when asked about their goal in life, would say, 'I want to be the next Abdul Kalam of India!'

Conclusion

Kalam was an inspirational leader, an intuitive and far-sighted technology innovator, poet and author, music lover and hard worker. He seemed to have effortlessly combined all these traits and promoted national interest with tireless devotion and fervour. As Prime Minister Modi put it:

> Kalamji saw poetry in a tree and energy that could be harvested in water, wind, and sun. We should learn to look at the world through his eyes, and with the same missionary zeal.

Abdul Kalam passed away on 27 July 2015, doing what he

liked best – teaching children. This Karma Yogi was at work and on his feet till the last moment. We are indeed poorer by the loss of this noble and humane personality, who was in our midst, goading us with his gentle persuasiveness towards higher goals in life. I hope this small tribute of mine will help in perpetuating his memory and inspire us to follow his example of selfless hard work and in some capacity serve the society.

1. APJ Abdul Kalam and his family 2. Dr APJ Abdul Kalam
3. Kalam at a lecture in ISRO

SECTION B

STATESMEN

INDIRA GANDHI
The First Among Pioneers

Sitakanta Mishra

AMONG THE DOMESTIC variables of nuclear weapon decision making, political leaders constitute the most vibrant institutional element. While articulating popular interests, political leaders as public representatives and policy-makers involve themselves in authoritative formulations on vital issues in the pursuit of national interest. In matters of nuclear weapon decisions, national political leadership is paramount, be it in a democratic setup or an authoritarian one. However, all decisions arrived at across the board are shrouded in mystery.

In India, the demand and decision to opt for nuclear weapons did not start with Indira Gandhi, who is otherwise regarded as the harbinger of the nuclear weapons era in India. India's nuclear weapons' journey started from 1950s itself when India decided to lay down nuclear research infrastructure with the declared intention to use atomic energy for peaceful purposes, leading to the 1960s when certain sections of the

Indian public started demanding for nuclear weapons. Prior to Indira Gandhi's decision to test in the 1970s, Lal Bahadur Shastri is credited for approving the Subterranean Nuclear Explosion Project (SNEP) as the first milestone in India's nuclear weapons journey. However, Indira Gandhi's role and position in India's nuclear arming choice cannot be relegated to the fringe. Her role and actions undoubtedly place her as the *primus inter pares* among the pioneers of India's nuclear weapons discourse. This chapter delves into the normative nuclear arming choice of India to evaluate Indira Gandhi's role in India's nuclear weapons discourse.

Indira Gandhi's emergence

Before examining Indira Gandhi's nuclear weapons related decisions, leading to the nuclear test in 1974, it would be appropriate to understand her personality traits as a decision maker, her rise to the helm of affairs and her attitude towards nuclear issues. At the outset, as the daughter of India's first Prime Minister, Indira Gandhi was undoubtedly exposed to national and international politics, and was imbibed with first-hand understanding of domestic as well as global power equations. During her father's premiership (1947-1964), Indira Gandhi was a key assistant and accompanied him on his numerous foreign engagements.

Indira Gandhi's political career started modestly in 1955 when she became a member of the Congress Working Committee (CWC) and Central Election Committee of the Party. She was the Chairperson of the National Integration

Council of All India Congress Committee (AICC), President of All India Youth Congress (AIYC) and Women's Department of AICC in 1956. In 1958, she was appointed as a member of the Central Parliamentary Board of Congress. As early as 1959, she was also elected as the president of the Indian National Congress and served till 1960, and subsequently in 1978. But the possibility of Indira Gandhi 'emerging as a long-term Indian political leader at the time was not apparent to everyone or perhaps many' (Prahladan, 2016).

Upon the death of Nehru in 1964, Indira Gandhi was appointed as a member of the Rajya Sabha and joined Prime Minister Lal Bahadur Shastri's cabinet as Minister of Information and Broadcasting. With the sudden death of Shastri in January 1966, she contested in the Congress Party's parliamentary leadership election and became the Prime Minister, defeating her rival Morarji Desai. It may be true, as Sudipta Kaviraj says, that 'the greatest qualification of Indira Gandhi at the time of her accession was…the fact that she was not too strongly associated with any policy line to give offence to any of the groups which dominated the polycentric structure of the Congress Party after Nehru's death' (Kaviraj, 1986: 1697). But, Kaviraj further says, 'Indira Gandhi came to power because she appeared to have a set of paradoxical political qualifications, most significantly, of indistinctness and ambiguity' (Ibid.).

This may have been the case in the beginning of Indira Gandhi's political career, however, it seems to have completely altered in later years with the quality of personal decisiveness in political-strategic matters. She is known to be a complex woman, tough politician and one of the greatest leaders in Indian history. Indira Gandhi was the first and only woman Prime Minister of India who is fondly remembered as the 'Iron Lady of India'. Besides her own style of decision-making, Indira Gandhi was assisted by eminent bureaucrats such as PN Haksar, DP Dhar, and BG Verghese, and influential intellectuals like NA Palkhivala, Karan Singh and others. However, most of them parted from her largely because they could not come to terms with her misuse of power and imposition of Emergency. During her Prime Ministership of sixteen years, some of her decisions were wise, some others were unwise, but undoubtedly, they are watersheds in India's independent history which changed India forever. One among them was her decision to conduct the nuclear test in 1974 which placed India on another side of the global nuclear divide.

The debate

The answer to the question, 'Why did India develop its nuclear weapon capability when it did and the way it did?' may lie somewhere in the politico-security spectrum. The political-nuclear interface is an established phenomenon, and in India, the nuclear weapon undoubtedly has an 'awesome appearance on the political scene' (Malcolmson, 1985: ix). Contemporary literature on nuclear weapons discourse seems to focus mainly on their strategy, deployment, number, quality and who possess them and how. It is to be understood that the policies on deployment, number, quality and possession are merely the ultimate stage in a process in which the decision may have been taken long ago (McLean, 1986: xiii-xiv). But to investigate why a nation arrives at a specific policy decision, it is necessary to unravel the complex processes of policy-making and to understand the variables that condition these processes. This is possible by undertaking an enquiry of the factors and the structures – the organisations, institutions and individuals or groups – which shape or control the process.

Raj Chengappa, in *Weapons of Peace: The Secret Story of India's Quest to be a Nuclear Power* (2000), asserts that, for long, India's nuclear issue has remained confined to a small, incestuous circle of strategic experts. Also, much of India's true nuclear history remains oral, but the nuclear questions straddle almost everything fundamental to India as a nation. A coherent authoritative official document on India's nuclear history is not available, and therefore, there exists considerable dissension and misperception. There is a handful of literature on the 'attitude and motivation' of individual political leaders, their ideology, personality and their motivation towards India's nuclear policy. Timothy D Hoyt in a paper 'Shades of Realism: The Impact of Personality, Ideology, and Systemic Pressure on Indian Foreign Policy' (2003) examines the role of ideas, individuals, organisations and institutions within the state and consequent structural pressures to explain state behaviour.

But an objective evaluation of the trajectory of India's nuclear discourse reveals that there is an element of 'circularity' and 'ad hoc-ism' to which Ashok Kapur in the book *India's Nuclear Option: Atomic Diplomacy and Decision Making* (1976) says the 'scientific and political czarism' premised on the attitudinal factors of the politico-bureaucratic community. His study directs attention to a need to investigate the interaction between politicians and scientists and other advisors in and outside the government. Perhaps keeping this aspect in mind, Praful Bidwai and Achin Vanaik in the book *South Asia on a Shot Fuse: Nuclear Politics and the Future of Global Disarmament* (1999) say, the crucial explanations for India going nuclear must be sought on the domestic terrain of Indian politics. Therefore, deconstructing the prevailing political context in which political decisions on nuclear weapons in India are undertaken and evolved would help. However, which political leader should be given credit and how much, in shaping India's nuclear weapons policy, is inconclusive. Resolve of the political leadership and understanding, coupled with geostrategic setting, has largely prompted nuclear weapons decisions. When one looks at the geostrategic situation during the 1960s and 70s, specific events like Chinese aggression, its repeated nuclear tests, America's nuclear intimidation during the Bangladesh Liberation War, etcetera have propelled the ruling political leaders to think in favour of the nuclear weapons programme.

The lessons of the Bangladesh Liberation War are especially unforgettable for India. The presence of *USS Enterprise* (the US nuclear-powered aircraft carrier) in the Indian Ocean during the war brought home the impression that India should go nuclear (Dixit, 2002: 213). Perhaps US pressure during the war, mingled with Pakistan's post-war plan to carry forward its nuclear weapon programme (in order to not face any humiliation by India), prompted Indira Gandhi to give green signal to the nuclear programme. India won her strategic goals under coercion from the two nuclear powers – US and China – and with the support of USSR (Ibid.). It is not only the external factors that prompted her to go for the test. Ashok Kapur says that 'the final decision [to test] was taken in the context of Mrs Gandhi's domestic troubles' (Kapur, 1976: 165). Further he says that, Indira Gandhi had not consulted any of her high ranking aids. India's then Defence Minister Jagjivan Ram, the second most powerful member of the cabinet, was informed on 8 May 1974 about the scheduled test (just ten days before the test), but he was not consulted for the decision. India's then Foreign Minister, Swaran Singh, was informed only two days before the detonation, thus indicating that the Ministry of External Affairs was totally bypassed (Ibid.).

In retrospect

Prime Minister Lal Bahadur Shastri died when he was just starting to gain popularity and control over the Congress Party. His sudden demise at Tashkent on 11 January 1966 brought Indira Gandhi to the helm of the Congress Party. By the time Indira Gandhi became Prime Minister, the split between 'pro-bomb' and 'no-bomb' within the Congress Party was wide. Indira Gandhi's rival for the Prime Ministership, Morarji Desai, Deputy Prime Minister then, was a staunch supporter of the no-bomb policy. Whereas KC Pant was vociferously arguing in favour of the bomb. But Indira Gandhi was more focussed on stabilisation of her leadership in the Party and her government. Therefore, like Shastri, she also avoided the liabilities of either embracing nuclear weapons or rejecting the option completely, mainly because of her unstable position in the Party. On 1 March 1966, when asked in the Lok Sabha whether the government would consider nuclear umbrella from other powers as an alternative to manufacturing nuclear weapons, Indira Gandhi said, 'the government's policy on this matter has been clear and it remains the same' (Quoted in Mirchandani, 1968: 43). She further said:

> We are anxious not to do anything which will precipitate the crisis and lead to the development of nuclear weapons in many more countries. The policy of restraint which we have adopted must, therefore, continue. This is not because we believe that certain big powers should have a monopoly of these destructive weapons, but because we are generally anxious to see total nuclear disarmament.
>
> (Quoted in Mirchandani, 1968:43)

Similarly, in the Rajya Sabha, replying to the question on how the government proposed to safeguard the nation's security in view of reports that Pakistan planned to test an atom bomb in 1968, Indira Gandhi said, Pakistan is not in a position to make a bomb, although it could conceivably get one from somewhere. She categorically stated:

> [I]f we make one, Pakistan will definitely get a bomb from somewhere or the other... We are anxious not to do anything which will precipitate the crisis.
>
> (Mirchandani, 1968:43)

By then, discussions on the nuclear issue had become a regular agenda in all meetings of the Congress Party. When the Congress Party met in the beginning of May 1966, its General Secretary KC Pant, among others, is known to have said that he did not wish to hustle the government into manufacturing the atom bomb, but wanted the country to advance technologically to a position, where if need arose, it could manufacture one quickly.

On 9 May 1966, China exploded its third bomb and claimed that it contained 'thermo-nuclear material', indicating that it had set off its first hydrogen bomb. As was expected, the news of China's test caused a storm in the Indian Parliament the next day. A leading member of the Praja Socialist Party, HV Kamath, called upon the Minister of External Affairs to make a statement on a 'matter of urgent public importance' (Mirchandani, 1968:44). The Congress members seemed to vie with opposition critics in deriding what they characterised as adherence to a moralistic stand (Ibid.). On the occasion, then External Affairs Minister, Swaran Singh said:

> We had made a careful assessment of the situation in consultation with our Service Chiefs and atomic energy experts even when the first nuclear device was exploded by China. ...the mere fact that China has carried out its third nuclear explosion does not vitiate the earlier conclusion, though at the same time, the policy is kept under constant review.

He further added:

> In the meantime, in the matter of peaceful development of atomic energy, we are pushing ahead and giving it top priority. The world recognises that we are one of the countries which are capable of becoming an atomic power in a reasonably short time.... Finally, I would assure the House that the defence of our territorial integrity will be the paramount consideration guiding our policy in this field.
>
> (Quoted in Mirchandani, 1968:45-46)

Prime Minister Indira Gandhi, intervening in the discussion, said:

> I do not think our policy is at all a negative one.... We are building up our atomic power. Of course, we are using it for peaceful purposes, but in the meantime we are increasing our know-how and other competence. I fail to understand myself how our production of one bomb or two bombs will help us. The belief that China can attack any country with nuclear weapons with impunity, I think, is a misconceived one.
>
> (Ibid.)

The above statements by two Congress Party stalwarts denoted a subtle change in the Government's nuclear policy. While some MPs viewed this policy as a continuation of Shastri's earlier statement, the media interpreted it as a definite shift. A day after the Lok Sabha discussion (10 May 1966), in the meeting of the Executive Committee of the Congress Party, the bomb issue was discussed for ninety minutes. Against the demands of various members for manufacturing nuclear weapons, Prime Minister Indira Gandhi gave no promise to reconsider the Government's nuclear weapons policy. On 21 May 1966, the All India Congress Committee (AICC) held a three-day session at Bombay. But the meeting discussed domestic issues only in spite of the opposition party's demand for a change in the Government's nuclear stand (Varma et al. 1968:364).

However, the Secretary of the Congress Parliamentary Party, KC Pant, urged the Government at a public meeting to give up its policy of 'nuclear celibacy' and make the country technologically prepared for making the bomb (Mirchandani, 1968:48). KC Pant claimed that China's prestige and status had risen after its entry into the Nuclear Club and criticised the Government for not making the bomb even under unfavourable circumstances. Equally, there were supporters of the no-bomb stand within the Congress Party. Morarji Desai, the then Deputy Prime Minister, was a staunch supporter of the Government's policy. On 21 August 1966, while inaugurating the Hiroshima Day Exhibition in New Delhi, he said, 'we must not be tempted, bullied or frightened into giving up that most sacred principle' (Ibid: 49). His stand didn't change even after China conducted its fifth nuclear test. He emphasised on rather strengthening India's conventional forces to help win any war with China.

Indira's decisions

By the time Indira Gandhi took over the Congress Party leadership, the debate over India going nuclear had become a recurring feature of domestic politics. The debate started to come to the fore during the Shastri era and got manifested, though late, during Indira's Prime Ministership. But there were two apparent trends that marked the nuclear weapons policy of India which continued till the late-1990s. First, as Perkovich quotes:

> [B]omb did not entail a specific decision in time but rather a continuous accretion of scientific and technical capability and political momentum, stymied occasionally by countervailing political, moral and economic considerations.
>
> (Perkovich, 2002:146)

Second, owing to perceived political cost and consequent hesitation for taking a straightforward nuclear weapons decision both by Shastri and Indira Gandhi, India's nuclear weapons policy became 'ambiguous'. Shastri and Indira Gandhi both seem to have avoided 'the liabilities of either embracing nuclear weapons or rejecting the option completely' (Ibid: 124).

During this time, according to Perkovich, 'three camps' were clearly visible in India's nuclear weapons debate (Perkovich, 1999:156). While one camp argued against nuclear weapons, the second proposed a crash programme to build a small nuclear arsenal to be delivered by aircraft, and in time, by ballistic missile. The third camp favoured a steadier, less provocative long-term arsenal with attendant command, control and warning infrastructure. But the Congress government's policy on opting for nuclear weapons was as usual based upon the socio-economic issues and peaceful use. At the Third

Conference of the Heads of State or Government of the Non-aligned Movement at Lusaka on 8-10 September 1970, Indira Gandhi highlighted the tremendous potential of the peaceful uses of nuclear energy, including peaceful nuclear explosions, for the developing world (*Foreign Affairs Reports*, 1970:85). However, according to K Subrahmanyam, 'there could be no question that the basic motives that triggered off the Indian nuclear test in 1974 were the China factor (and US nuclear intimidation)' (Subrahmanyam, 1998:32). If one examines China's activities on the nuclear front during the late sixties and early seventies, Subrahmanyam's assertion seems to hold truth. By that time, there was sufficient evidence regarding China devoting substantial resources to the development of both nuclear warheads and missile delivery systems. Were all the Chinese preparations meant for India?

By then, the view among the strategic community was that China's professed purpose in going nuclear was to 'oppose the nuclear blackmail and threats by US imperialism and its collaborators and to oppose US–Soviet collusion for maintaining nuclear monopoly'. Even in a communiqué issued after the third nuclear explosion, the Chinese Government said:

> China's sole purpose in developing nuclear weapons is defence and her ultimate aim is to eliminate the nuclear weapons.... At no time and in no circumstances will China be the first to use nuclear weapons.
>
> (Mirchandani, 1968:93)

Also, within the Indian intellectual circles, there was a view that if an aggressive nuclear threat is aimed against India, there would certainly be a counter threat from the major nuclear powers (Ibid.).

Irrespective of all solemn declarations by China, India was definitely apprehensive of China. The major question was, will the United States risk a nuclear attack on its own cities and use nuclear weapons against China to discharge moral obligation to defend India? At that stage, India had the Canada-India Reactor and a stockpile of weapon-grade plutonium was being built up. With this plutonium, Bhabha Atomic Research Centre (BARC) had designed the Purnima reactor and this enabled the scientists to gather data necessary to design a nuclear explosive device. Since, there was commitment that the Canada-India Reactor would not be used for purposes other than peaceful, finally Indira Gandhi gave the go-ahead to the scientists sometime in October 1972 for a 'peaceful nuclear explosion' (PNE) but no record is available to explain her decision (Subrahmanyam, 1998:30).

To decipher the motives of Indira Gandhi's decisions, particularly on security and nuclear matters, one needs to examine the relationship among the following factors as identified by Ashok Kapur:

- Indira Gandhi's motivation, political interests and compulsions;
- assessments by the Indian bureaucracy of external and internal security threats and the budgetary, military and intelligence requirements; and

- motivations and policies of external powers that impinged upon India's security concerns.

(Kapur, 1976:49)

To appraise Indira Gandhi's nuclear decisions, one needs to first see Indira Gandhi as an actor in the theatre of Indian politics. During the initial stage of her Prime Ministership, her compulsion was to create a string of favourable effects and temporary success to strengthen her position (Chakma, 2005). The slogan '*garibi hatao*' helped her to win the election with absolute majority. According to Ashok Kapur:

> Her January 1966 decision to cancel the nuclear test project may be attributed to inexperience to the pressure of other priorities (i.e., staying in power). Her decision in February 1968 not to sign the NPT reflects fear of Cabinet disapproval and the nature of bureaucratic advice. It seems, during the whole of the late-1960s, neither she understood nor cared for nuclear affairs. Yet, she was making the case against the NPT on grounds of principle and national interest.
>
> (Kapur, 1976: 49)

Also Kapur views Indira Gandhi's 1974 Pokhran test decision as 'a product of scientific advice, a diplomatic demonstration to the major powers in India's environment, a popular gesture and a method to disarm India's bomb lobby'. But, it is because of this bureaucratic advice, external pressure and her unfamiliarity with the discourse, that she failed to pursue the test to its logical conclusion (Kapur, 1976:52).

Therefore, it seems, Indira Gandhi's compulsion was not to formulate policies but to engage in *ad hoc* policy actions primarily to promote her ascendancy in Indian politics and to take advantage of the immediate situation.

The run up to the 1974 Pokhran-I tests was marked by many regional security problems which Indira Gandhi had to cope with. Remarkable among them was the 1971 Bangladesh crisis. Also, external powers, especially the USA, former USSR and China played their part influencing the South Asian security environment. But, there are differing views on whether the Bangladesh crisis and consequent involvement of the major powers especially USA by sending the USS Enterprise to Bay of Bengal was a factor in the Congress Government's (Indira Gandhi's) decision to go for the nuclear test. The common argument is that the security threat arising out of the Sino-US rapprochement and the US presence in the Indian Ocean during the crisis was a determining factor in Indira Gandhi's decision for the 1974 tests, which George Perkovich vehemently discounts. Perkovich asserts that 'the decisive factors that led to the detonation of a nuclear device in 1974 were the Atomic Energy Commission's determination to prove its mettle and Indira Gandhi's sense that India would gain confidence in itself as a nation and her as a leader' (Perkovich, 2002: 161).

According to Perkovich, 'the final decision to conduct the test was the result of an ad hoc, intuitive process that lacked rigorous security analysis' (Ibid: 161). He views that the Pakistani nuclear capability was incipient at that time, and the war indicated that 'China did not seek militarily to threaten India's security or challenge its pre-eminence in the subcontinent' and that no indication is there that China played any direct role in assisting Pakistan in the crisis (Ibid: 165). What then prompted the Congress Government, led by Indira Gandhi, to decide on the test? According to a survey, 46 per cent of the members of Congress Party were in favour of going nuclear, while 49 per cent were against it. The rest were undecided (Nanda, 1971:39).

One can argue that the super power threat to India was more visible during Indira Gandhi's leadership. First, during the East Pakistan crisis, the USA, China and USSR were all anxiously watching the unfolding of events, and in many ways indicated their seriousness to interfere if required. In fact, non-aligned India probably had faced the gravest threat to her sovereignty during the crisis. After his secret visit to China, in July 1971, Kissinger told the Indian Ambassador to Washington, LK Jha, that in case of an Indo–Pakistan War over the Bangladesh issue, if the Chinese were to intervene on behalf of Pakistan, then the US would not be in a position to support India (Subrahmanyam, 1998:30). According to K Subrahmanyam, the event was 'clearly an attempt at intimidation, to prevent India from acting to defend its own national interests and security and conveying a veiled threat about the possible intervention by a nuclear China in support of Pakistan' (Ibid.). In fact, later it came to light how Kissinger pressed the Chinese to intervene against India and the Chinese refused to do so fearing Soviet reaction (Kissinger, 1999).

Subsequently, when India intervened in the crisis that

resulted in emergence of Bangladesh as a sovereign nation, USA sent in its Task Force 74, headed by the nuclear aircraft carrier *USS Enterprise*, on board of which nuclear weapons were then standard equipment. But, later it was revealed that there were no specific operational directions to the Enterprise mission (Subrahmanyam, 1998:31). President Nixon disclosed later that he did contemplate the use of nuclear weapons at that stage. But the event was sufficient for the Indian government 'to assume the worst and treat it as an act of nuclear intimidation' (Ibid: 30.) It is assumed that this experience of nuclear intimidation must have influenced Indira Gandhi in giving the green signal to the Atomic Energy Department to go ahead with the nuclear tests in 1972 (Ibid: 30). It is reported how Indira Gandhi apprehended the US as a direct threat to her political career. The accounts by Inder Malhotra and Pupul Jayakar reveal that Indira Gandhi feared a covert Western attempt (through the Central Investigation Agency) to oust her from power in the mid-1970s, particularly in the aftermath of the violent coup that eliminated Sheikh Mujibur Rahman of Bangladesh (Jayakar, 1992; Malhotra, 1990).

Archival studies reveal that there may be a link between the CIA activities in Chile and Indira Gandhi's determination to go in for nuclear weapons and means of delivery. Debak Das, in his archival study on Daniel Patrick Moynihan's (former US Ambassador to New Delhi) leaked cable to the State Department titled, 'The United States as a Counter-Revolutionary power – The Case of India and Chile' reveals how Indira Gandhi was apprehensive of US intentions in the wake of the bloody removal of Chilean President Allende (Das, 2019). When the CIA director William E Colby's secret House Armed Services subcommittee testimony was leaked, the role of the United States in the bloody removal of President Allende was confirmed. In a speech in Madras on 9 September 1974, Indira Gandhi reportedly said that 'reactionary forces, helped from outside, were using the current difficulties to whip up the people against the government and in this context it had become necessary to make India strong….' In his report, Moynihan condemned the CIA's covert operations in Chile, believing that such actions would lead Indira Gandhi to 'proceed to develop nuclear weapons… preaching nonviolence all the way'. Ambassador Moynihan stated that:

> She [Gandhi] knows full well that we have done our share and more of bloody and dishonourable deeds. This as such is not her concern. She knows all too much of such matters. It is precisely because she is not innocent, not squeamish and not a moralizer that her concern about American intentions is real and immediate….

Moynihan continued:

> …Do not think, fellow Americans, of beguiling Indira Gandhi with talk of cultural exchange, joint industrial undertakings or a few shiploads of cheap food. Her concern is not economic. It is political. Nothing will change here till she is satisfied that the United States accepts her India. She does not now think we do. She thinks we are a profoundly selfish and cynical counterrevolutionary power. She will accordingly

proceed to develop nuclear weapons and a missile delivery system, preaching non-violence all the way.

Owing to all these factors, perhaps Indira Gandhi was clear about whom to deter primarily with a nuclear arsenal – China and, after 1971, more remotely, the US (Karnad, 2002:306). In subsequent deliberations of the Congress Party, though the debate on the nuclear threat to India's national security was not focused on any particular nation, pronouncements by the Congress Party members in public forums often vindicated that one of the motives behind India acquiring nuclear weapons was the perceived threat from super power activities. For example, the Congress leader and defence minister, C Subramaniam, while delivering an address at the National Defence College in 1979, spelled out seven factors that might provoke India into exercising the weapons-option. Out of the seven reasons he spelt out, six involved major powers (Subramaniam, 1979:12). They include,

1. Pakistan acquiring nuclear weapons;
2. no halt to the growth of 'warheads' in the American and Soviet nuclear arsenals;
3. the UK, France and China continuing with their nuclear weapons programmes;
4. the attitude to clandestine nuclear weapons development projects in Israel, South Africa and Taiwan;
5. the type of access offered to non-nuclear US allies (NATO members) and Warsaw Pact partners (Soviet allies) to the American and Soviet nuclear arsenals (in other words, nuclear

umbrella provided to allies of super powers);

6. the interventionist tendencies of the nuclear powers; and
7. the process of legitimation or de-legitimation of nuclear weapons.

Another Congress leader who later became President of India, Shankar Dayal Sharma, raising concerns over the military bases in Indian Ocean, moved a resolution in the AICC meeting in December 1980 saying:

> As you are aware we have to worry about Indian Ocean. We had passed a resolution in the last Session that Indian Ocean should be made a zone of peace but we find that not only a military base in Diego Garcia is being established but also it is being expanded. In 1971, India succeeded in getting a resolution passed in UN that it should be a zone of peace but in spite of that we find they are going merrily about it.
>
> (Minutes of the AICC, December 6-7, 1980)

During early 1972, the nuclear policy debate resurfaced at the domestic level. On 17 March 1972, in the Lok Sabha, the government was asked whether the war with Pakistan in 1971 'brought out the necessity for India to have more modern weapons' and if so 'whether government propose[s] to embark upon the manufacture of nuclear bombs' (*Lok Sabha Debates*, 17 March 1972). Defence Minister Jagjivan Ram said, 'modernisation of army is undergoing, and nuclear energy is meant only for peaceful purposes, and possession of nuclear weapons is not a substitute for any other military preparedness'

(Ibid.). In the AICC meeting at New Delhi on 1-2 June 1972, the Congress Party led by Indira Gandhi adopted the resolution saying:

> [P]eople all the world over were convinced that atomic weapons should neither be produced nor used. The only way to save the world from the catastrophe of another world war was acceptance of co-existence.
>
> (Zaidi, 1984: 162–63)

The Party endorsing the Prime Minister's initiative in holding talks with Pakistan in the quest for durable peace in the subcontinent, Congress leader and then Foreign Minister, S Swaran Singh, said, 'it is our earnest hope that the...two neighbours – Pakistan and China – would also move in the direction in which there may be good neighbourly relations....' (Ibid: 168).

Pressure continued to build within the Parliament for a more robust nuclear posture and during the debate over the defence ministry budget, all parties, except the two communist parties, called for developing nuclear weapons. This debate resurfaced in the Parliament in November 1972. Answering a pointed question on the 'progress...in feasibility study and other preparations for the experimental nuclear explosions for peaceful purposes,' Indira Gandhi said:

> [T]he Atomic Energy Commission is constantly reviewing the progress in the technology of underground nuclear explosions both from the theoretical and experimental angles and also taking into account their potential economic benefits and

possible environmental hazards.

(Lok Sabha Debates, 3 November 1972: Col. 49)

The above statement by Indira Gandhi makes it clear that the possibility to prepare for an underground test was already there by that time. Perkovich, by citing many factual evidences about the timing of the decision for underground testing, asserts that the decision was taken prior to the war with Pakistan and the US deployment of the *Enterprise* (Perkovich, 2002:171–72). He cites Jagjivan Ram saying that the decision was taken three years before the test, placing it somewhere in 1971. But his interview with Homi Sethna reveals a different picture – Indira Gandhi seems to have 'authorised fabrication of a device for peaceful nuclear explosion on 7 September 1972'. However, Perkovich writes:

> [I]t may be conjectured that support in principle for developing a nuclear explosive device was solidified by late 1971, that concentrated work on building the vital components began in spring 1972 and that formal prime ministerial approval to make final preparations for PNE occurred in September 1972.
>
> (Perkovich, 2002: 172)

But the decision or the preparation was never recorded or made public as one can deduce from the Rajya Sabha debate on 15 November 1973, when Indira Gandhi was asked in writing 'whether any final decision has been taken to conduct experiments to develop nuclear blast technology... whether any time schedule has been set for the purpose', to which Indira Gandhi replied in the negative saying that such decisions would be taken 'only after satisfactory answers to all these problems [theoretical and experimental questions; economic questions; environmental and ecological effects] are available' (*Parliamentary Debates*, Rajya Sabha, 15 November, 1973: cols. 138–39).

In contrast to Perkovich's argument, for K Subrahmanyam, 'there could be no question that the basic factors that triggered off the Indian nuclear test in 1974 were the China factor and US nuclear intimidation' (Subrahmanyam, 1998:32).

The issue of China as a factor in India's 1974 nuclear tests is an open-ended debate. But from Brajesh Mishra's account, it seems there was considerable concern on China's behaviour vis-à-vis the 1971 war, at least at the official level. Probably, that is why Indira Gandhi was convinced about Brajesh Mishra's proposal to engage China. But, how far China factored in her decision for the nuclear test in 1974 is still shrouded in mystery. It is not clear whether regional security imperatives were behind the decision for the nuclear test in 1974, but generally it is viewed that the decision was based partly on:

> [I]ntuition, bureaucratic and technological momentum, and the personal, enigmatic calculations of the politically beleaguered prime minister under the influence of trusted nuclear scientists.
>
> (Perkovich, 2002:178)

Most Indian and foreign scholars are of the view that

political considerations moved Indira Gandhi to decide for the nuclear test. It is true that the political situation of the country was worsening in 1974, even though the Congress Party won narrowly in the Uttar Pradesh state election in February. In January 1974, Morarji Desai, Indira Gandhi's rival, erupted in revolt and announced his 'fast unto death' to compel her to dissolve the Gujarat state legislative assembly (Frankel, 1978:526-27). Equally in Bihar, JP Narayan was protesting in demand for 'all round devolution, political, economic, social, educational, moral and cultural' (Frankel, 1979:532). Also, political turmoil spread across the nation when 1.7 million railway workers called for a strike that began on 8 May, and Narayan was reportedly coordinating the activity. According to the Congress Party report, the Right and the Left appeared to be attacking all government policies (AICC Resolution, 1974: 16). Definitely the political situation of the country was precarious.

However, the scientific community, who were involved in the test preparations, especially Raja Ramanna and PK Iyengar, were of the opinion that the PNE was simply done when the scientists were ready (Ramanna, 1991:89). At the July 1974 session of the Congress Working Committee (CWC) in New Delhi, a resolution was adopted by the Party hailing 'the achievement of Indian scientists in carrying out successfully the underground nuclear experiment for peaceful purposes....' The AICC, congratulating the scientists and the Prime Minister for this achievement, welcomed 'the clear statement of the Government of India reaffirming their intention of using nuclear technology for peaceful purposes and not for manufacturing nuclear weapons' (Zaidi, 1984: 219). The Congress Working Committee rejoiced in the fact that 'this notable and important experiment has been based on indigenous effort,' and that 'in the process, India has not violated any of its international obligations' (Ibid.).

If the decision was purely a scientific imperative, Indira Gandhi could have taken her cabinet into confidence, at least her close aides, for such an important event. Important members of the Cabinet like the Defence Minister Jagjivan Ram, the Minister of External Affairs Swaran Singh and others were informed only a few days before the explosion. This may have been done to maintain secrecy, but it shows that the reasons for the 1974 nuclear test decision lie not only in the domestic politics and regional security considerations, but the decision-making style of an individual decision-maker. Indira Gandhi's decisions on future test proposals made by the scientific community clearly mark this aspect. For example, by 1982, BARC and DRDO, under the leadership of Ramanna and VS Arunachalam respectively, 'produced a design for a device that would weigh between 170 and 200 kilograms' and in early 1983 they made their case for another nuclear explosion to Prime Minister Gandhi (Perkovich, 2002: 242). During their meeting, Ramanna and Arunachalam tried to convince Indira Gandhi on whether India should or should not acquire the bomb before Pakistan does (Gupta and Pabian, 1997: 108-188). As the scientists wished, she tentatively approved the request. But within twenty-four hours, she reversed her decision and the scientists were instructed to hold on. Even during the early years of her second tenure as Prime Minister, she was not in favour of forswearing further nuclear explosions. (Ibid.) However, no information is available regarding what factor(s) actually changed her mind.

In the context of regional or national security, during this period, India perceived a number of threats, especially emanating from Pakistan. First, by this time it was clear that Pakistan was nuclear capable. AQ Khan himself was boasting of leading his country ahead of India in nuclear matters (Sreedhar, 1987: 57). Second, there were reports on Sino-Pakistan nuclear collaboration. Thirdly, there was fear of Pakistani support to Sikh insurgency in Punjab. At best, Indira Gandhi, in her second term, reversed Morarji Desai's nuclear aversion policy by not foreclosing the possibility of further nuclear explosions. But throughout 1982, she downplayed the nuclear issues. For example, during her meeting with US President Reagan, she categorically denied any intention to conduct a test, instead told him that the test rumour was generated by the Bharatiya Janata Party (BJP) (Delhi Domestic News Service, 1982). In an interview in January 1982, Indira Gandhi, while replying to whether India would follow suit if Pakistan developed nuclear weapons, said:

> The threat is one of conventional arms and not of nuclear weapons. I am very much against war of any kind and especially, nuclear war. But I don't think I can give a categorical answer and bind the Indian government to any position.
>
> (Statement on Foreign Policy, 1982: 25–26)

Also, Prime Minister Indira Gandhi vehemently rejected the US intelligence reports on Indian intention for launching air strikes against Pakistan's uranium enrichment plant at Kahuta and the reprocessing facility at Pakistan Institute of Nuclear Science and Technology (PINSTECH) in Rawalpindi, which according to the report was planned in March 1982. Neither did she authorize such an attack, nor permit to conduct the second nuclear explosion despite technical readiness. One factor could be the domestic political situation that demanded more attention, as in the State Assembly election that year, the Congress Party had faced major defeat in Andhra Pradesh and Karnataka (*India Today*, 1983:18). Second, Indira Gandhi's curiosity for nuclear explosion might have waned owing to international pressure after the 1974 tests. Third, since Pakistan did have nuclear arsenals by that time, she might have thought that it is not prudent to conduct a test that would create a chain reaction from Islamabad, further deteriorating regional and national security. Until July that year, Indira Gandhi was categorical that 'India had no intention of embarking on the nuclear weapons programme' (*India News*, 1983:1).

Overall, during this period, Indira Gandhi's nuclear weapons policy was based on 'a less risky strategy of bide time, avoid provocations, pursue diplomacy and keep the nuclear powder dry while downplaying explicit interest in nuclear weaponry' (Perkovich, 2002:254). Of course, there were number of deliberations in the parliament on the issue of Pakistan's nuclear programme and the Sino-Pakistan nexus and India's response thereof.

Indira's second term

During Indira Gandhi's second term as Prime Minister, the 'Nuclear-Pakistan' factor was not enough to budge Congress Party's ambiguous nuclear weapons policy. However, this does not mean that the Indira Government was callous about Pakistan. While addressing a gathering of Indian Army Commanders on 11 October 1984, Indira Gandhi said that 'Pakistan's nuclear programme has brought about a qualitative change in our security environment' (Oberdorfer, 1984). On 16 October 1984, Indira Gandhi ordered to seal the Indo–Pak border (Ibid.). But, it is unclear why she became more calculative this time even though the security situation was warranting some action in the nuclear front than before, and more importantly, when the scientific community was prepared and in favour of such an act. But, Indira Gandhi refused a meeting with concerned scientists subsequently on this issue (Perkovich, 2002: 244). The probable reasons for Indira Gandhi's attitude towards the nuclear issues may be:

- the negative consequences of 1974 test;
- her own abhorrence of nuclear weapons;
- absence of a Pakistani nuclear provocation in the form of test explosion in response to India's 1974 test, which might have given an impression that India is better off holding its nuclear fire (Ibid.). Her aides – PN Haksar, PC Alexander, Krishnaswamy Rao Sahib – have also brought her the probable international and regional repercussions resulting out of another nuclear test explosion; and
- the fear of assumed attacks on each other's nuclear facilities

vindicated that even without nuclear bombs a rudimentary deterrence is in operation.

Probably owing to these reasons, there emerged a buffer between the decision-maker and the 'strategic enclave' (scientists and other votaries of nuclear weapons). The people around Indira Gandhi might have persuaded her that nuclear weapons would not provide usable solutions to imminent problems as they are not usable and India's well-being requires other priorities. This also shows Indira Gandhi's 'chariness towards the nuclear scientists and the military, as well as the increased role of professionals, broad-gauged advisors in the Prime Minister's office....' (Ibid: 259).

According to General K Sundarji, this led to a situation between the mid-seventies and the mid-eighties, where 'India's nuclear decision-making appears to have enjoyed something between a drugged sleep and a deep post-prandial siesta' (Sunderji, 1993: xiv). Further, he asserts that India did not have 'a coherent nuclear weapons policy and worse still', she [Indira Gandhi] did not have 'an institutionalised system for analysing and throwing up policy options in this regard' (Ibid., p. xv). Most importantly, one tragic domestic event that altered India's nuclear trajectory was the Golden Temple calamity and consequent assassination of Indira Gandhi by her Sikh bodyguards. The Sikh uprising and Operation Blue Star deepened the Indian polity's preoccupation with the challenge of internal order and drove the national leadership's attention away from the nuclear issue (Kapur, 2009: 79). One can only speculate in which way India's quest for nuclear weapons could have turned, had Indira Gandhi remained alive.

One may conclude that the Congress Party did not link outright the imperative of India acquiring nuclear weapons for national security. Rather, the secretive nature of India's nuclear policy seems to have continued from Indira Gandhi's tenure. For that matter, the Congress Party maintained a no-bomb posture publicly while inching towards weapons capability covertly. In the aftermath of the East Pakistan crisis, though Indira Gandhi is known to have given a go ahead for feasibility study of a test, the Congress Party deliberations and public pronouncements maintained a no-bomb posture. So, the Party has never publicly established or drawn the link between national security scenario and the imperative of India acquiring nuclear weapons.

Dealing with non-proliferation and disarmament

During 1966 to 1968, concerns regarding nuclear non-proliferation were raised in the Indian Parliament several times. Each time the Congress Party leaders reiterated the Party's stand by saying that though India has the capacity to produce a nuclear bomb, it is in the larger interest that proliferation of the same technology should be curbed and New Delhi is 'in the front row' in such initiatives. On 8 March 1966, during the Question Hour in the Rajya Sabha, replying to US Dugal's question, 'whether Government's attention has been drawn to press reports about Pakistan's plans to test its first atom bomb', Indira Gandhi said:

> The Government of India deplores the conducting of nuclear tests, despite the Partial Test Ban Treaty (PTBT), as an act injurious to the welfare of mankind, and contrary to the efforts of all peace-loving nations to achieve a Comprehensive Test Ban.
>
> (Rajya Sabha Debate, 1969: 53)

Meanwhile, USA and Soviet Union developed common interests against horizontal nuclear proliferation, but presented separate drafts of the nuclear non-proliferation treaty. During a discussion on call-attention motion, on India's reaction to this agreement between both the super powers, in the Lok Sabha on 14 March 1968, Tarkeshwari Sinha asked about the exact position of India on this matter. Indira Gandhi replied:

> [T]he Government of India's position is not rigid. If the treaty were to be changed and we feel that it was in our interest, we shall sign it and we have made it clear in no uncertain terms that the present draft treaty does not give us any satisfaction.
>
> (Lok Sabha Debates, Phalguna 24, 1889 (Saka), 14 March 1968:739)

And Indira Gandhi also made it clear that:

> [T]he guarantee of security could not be a quid pro quo for signing the non-proliferation treaty in its present form.
>
> (Ibid.)

So, during 1967–68, the nuclear question in India 'shifted from whether India should actually produce nuclear weapons to whether India should sign a treaty relinquishing the right to produce weapons' (Perkovich, 1999:134). The non-proliferation issue came up in the Parliament during the voting on the Demands for Grants under the control of the Ministry of External Affairs on 3–5 April, 1968. Members like MR Masani and ML Sondhi from Communist Party of India (CPI) and Jan Sangh, respectively, raised the non-proliferation issue and India's policy thereof. The main contention of both these parties was that the 'Government seems to be altogether without a policy on this subject'. On 5 April 1968, replying to this debate in the Lok Sabha, Prime Minister Indira Gandhi gave a passionate long speech saying:

> ...we shall be guided entirely by our enlightened self-interest and the considerations of national security and of course, adherence to our values.... We have already made it clear that the draft treaty in its present form does not fully conform to the principles enunciated in the General Assembly Resolution No. 2028 of the 20th Session.
>
> ...It is very important, therefore, for the nuclear weapon powers to undertake as soon as possible meaningful negotiations on a series of measures leading to nuclear disarmament. The present draft treaty acknowledges the need for such negotiations, but unfortunately, the non-participation by some nuclear weapon powers will make it only partially effective and what is more, the other nuclear weapon powers insist on their right to continue to manufacture more nuclear weapons.
>
> (Lok Sabha Debates, 1968: 55–57)

In the same speech, Indira Gandhi stated:

> [T]he Government of India does not propose to manufacture nuclear weapons. This is a unilateral decision taken many years ago and is unrelated to the treaty on non-proliferation of nuclear weapons.
>
> (Ibid.)

Indira Gandhi promised to make efforts for nuclear disarmament because, in her view, 'it is only through nuclear disarmament that discrimination would be eliminated and equality between nations re-established' and this is in 'the best interests of the country'. At the same time, she also warned:

> [N]ot signing the treaty may bring the nation many difficulties. It may mean the stoppage of aid and the stoppage of help. ...I personally think that although it will involve sacrifice and hardship, it will be the first step towards building the real strength of this country and we will be able to go ahead on the road to self-sufficiency.
>
> (Ibid.)

However, Indira Gandhi's concerns on the consequences of not signing the treaty received sufficient support from the Congress Party members. Vijaya Lakshmi Pandit was of the following opinion:

> [T]he Government would not be high-pressurised into signing any treaty which is not in India's interest. There is no need to do that through fear…America cannot afford to deny aid to this country. If she wants to preserve democracy

> in India, she will give you aid whether you sign or you do not sign. It is for our government to take a strong stand on the matter.
>
> (Lok Sabha Debates, 5 April, 1968, Chaitra 16, 1890 (Saka): 2640)

But the discussions in the Parliament did not exhibit the subtle shift that the Congress government was in the process of embarking on in its stand on the Non-proliferation Treaty. On 27 March 1967, during the Lok Sabha debate on nuclear policy, Minister of External Affairs, MC Chagla unequivocally expressed that India should not sign the NPT as it stands, because, in addition to failing to provide security for non-aligned states such as India, it would impede future development of peaceful uses of nuclear energy. The Secretaries Committee expressed its hope that the treaty would be improved and that the issue of protection of non-nuclear states would be clarified. In consonance with his apprehensions, Chagla, in a press interview on 3 May 1967, characterised the draft treaty as discriminatory and a 'step away from disarmament'. He said that India could not give up the right to experiment with nuclear technology for peaceful purposes (Foreign Affairs Record 1967, MEA Library 1995: 101).

At that juncture, the Congress Government was buffeted between two opposite currents:

- for the first time, the proposed non-proliferation treaty signified the emergence of a new, yet undefined bond between the two super powers who had hitherto been rivals;

- in opposing their initiative on the non-proliferation front, New Delhi was completely going against the entire trend of foreign policy in the past.

(Mirchandani, 1968:146-47)

Despite the risk, the Congress government expressed its dismay over the treaty. Deputy Prime Minister Morarji Desai, during his foreign tour, conveyed to Robert McNamara India's strong objection to the treaty both on the grounds of security and discrimination. Defence Minister Swaran Singh formally pronounced in the UN General Assembly that nuclear disarmament could not be achieved by the preservation of exclusive rights, privileges and options 'sought to be retained by certain armed and powerful countries, while measures are to be taken to limit the actions of the threatened and unarmed countries' (Ibid: 148). Speaking at Warsaw during a state visit, Prime Minister Indira Gandhi said:

> We see danger both in horizontal and in vertical proliferation. While seeking agreement on the non-proliferation treaty, we should like to emphasise the balance of obligations which must prevail for nuclear disarmament. The stockpiling of armaments is a cause for fear and distrust.
>
> (Quoted in Mirchandani, 1968: 149)

The nuclear non-proliferation treaty and India's response remained a major political issue at home. Almost every political party and its members participated in the debate on what should be India's policy on the matter. But, interestingly,

Shyam Bhatia points out a different perspective when he depicts that it was a decisive time as Congress Government was actually preparing its justification for carrying out future nuclear test then (Bhatia, 1979: 141–42). According to him:

> [I]t was only in 1968, after four years of running the plutonium plant and after sufficient stocks of plutonium had accumulated to begin basic theoretical work on explosions, that Sarabhai approached Mrs Gandhi and told her preparations could now begin for a nuclear test, if the government so wished.
>
> (Bhatia, 1979:141).

Bhatia further says:

> [B]y the time the Purnima project had been sanctioned in 1968, the government had a wide range of reasons to justify a nuclear test. They included domestic pressures, security problems emanating from the Chinese nuclear programme, an unsatisfactory Nuclear Non-Proliferation Treaty and economic benefits associated with the development of nuclear explosive technology. In consultation with Atomic Energy Commission, Mrs Gandhi decided to base her strategy on the last reason, just as Shastri had done for reasons of cost effectiveness.
>
> (Ibid.)

But completely contrasting views can be found in Indira Gandhi's nuclear policy decisions subsequently. According to Ashok Kapur, 'the internal setting was such that Indira Gandhi was in the process of gaining her ascendancy in Indian politics in the late 1960s, and she neither understood nor cared for nuclear affairs' (Kapur, 1976: 52). Kapur suggests that one needs to see Indira Gandhi, indeed the Nehru family, as actors, and Indian politics as theatre where the aim is to create an illusion of reality and to achieve a rapport with the audience to achieve an effect of the moment. He cites the example of her January 1966 decision to cancel the nuclear test project which he attributes to Indira Gandhi's inexperience, to the pressure of other priorities (i.e. staying in power), and to the nature of bureaucratic advice. The reason behind her decision, in February 1968, to not sign the NPT, may be her fear of Cabinet disapproval, the nature of bureaucratic advice and the impression that the Treaty was not popular in India. Kapur considers that Indira Gandhi made the case against the NPT on grounds of 'principle and national interest' which are a vital part of her and India's nuclear posture on the treaty, but had nothing to do with the 'motivation' behind this decision. Rather, consideration of her own and her party's prestige, and pressure of bureaucratic politics from below, and not a carefully considered assessment of the national interest, governed her decision regarding NPT (Kapur, 1976:49). Kapur advances the hypothesis that during 1966-69, Indira Gandhi's priority was to establish her ascendancy in Indian politics. In other words, domestic politics was paramount in all her policy pronouncements including on nuclear matters.

Though Indira Gandhi, soon after becoming Prime Minister in 1966, launched a dynamic programme for nuclear research development, she consciously 'qualified it by restricting it exclusively for peaceful purposes' (Kaul, 1988:157). Though, domestic political considerations were paramount in Indira Gandhi's thinking, overall the Congress Party's stand on NPT reveals that she had the full confidence of her Party members. Indira Gandhi's final decision to reject the NPT was taken after the issue was extensively debated in her cabinet.

Interestingly, Ashok Kapur, in his earlier writing (*India's Nuclear Option*), says that 'in light of Indira Gandhi's preoccupation with domestic concerns, India came very close to signing the NPT, given the foregoing array of bureaucratic manoeuvres' (1976: 196). Many senior bureaucrats from the Ministry of External Affairs (except VC Trivedi and TN Kaul) like Rajeshwar Dayal, Jagat Mehta and Rikye Jaipal favoured signing the treaty. Morarji Desai and YB Chavan were the principal cabinet members who argued against the treaty. Kapur says the original draft of Indira Gandhi's speech in late November 1967 favoured India's adherence to the NPT. But her cabinet was unanimously against the treaty (ibid.). Therefore, her decision not to sign might not have been based on any deep conviction in principle against NPT; rather it was based on a careful reading of Indian public opinion polls which showed that 'a majority rejected the NPT and favoured an Indian decision towards nuclear weapons' (Ibid.). It shows that there was considerable ambivalence within the Congress Party regarding the decision on NPT.

It is possible that nuclear issues were among the least of Indira Gandhi's worries as her government was more concerned about the general elections of April 1967. How important the nuclear issues were for her can be perceived if one analyses her speech in the Lok Sabha on 5 April 1968. One can get the impression that Indira Gandhi and the Congress Party's policy on non-proliferation was nothing different from India's traditional vocal aversion to nuclear weapons. However, if the assumption 'political status of the decision-maker is paramount in shaping nuclear policy' is any guide, 'foreign policy issues have a domestic use' especially in the hands of the Prime Minister who does not trust, and is unwilling to give credit to his or her colleagues (Ibid: 197). Then, it is not merely on the basis of the merits of the issue (NPT) that the decision was settled in India. But, once the threat from the domestic politics front (adversaries) passes, public opinion does not become a key variable. This aspect is clearly perceptible if one sees how Indira Gandhi, after gaining a majority in the Parliament, went for the nuclear test in 1974 without taking into confidence other fellow party members. But Indira Gandhi, during her speech in Lok Sabha on 2 November 1966, mentioned that 'all parties, with the exception of Hon. Member Shri MR Masani, have generally supported the stand taken by the Government of India in not signing the Treaty in its present form' (Srivastava, 1985:34). This indicates that all other parties were of the same view as the Congress Party regarding the NPT. It also suggests that Indira Gandhi must have had extensive consultations with other party members before arriving at the decision to withhold India's support for the treaty.

Whatever may be the nuances of Indira Gandhi's decisions

on NPT, it is quite clear that the Congress Government led by her was firm to 'protect India's national interest by not unilaterally surrendering its sovereign right to manufacture nuclear weapons even though she has of her own free choice decided to exploit nuclear energy for peaceful purposes only' (Khilnani, 1984:42).

In his statement before Parliament on 17 March 1967, the Foreign Minister MC Chagla made it clear that 'the non-proliferation treaty should be such as to not impede the growth of nuclear science and technology in the developing countries where the need of such development is great'. Again, on 27 March 1967, he said in a statement to both houses of the Parliament that India should examine the text of any draft treaty submitted to the Committee in the light of the principles enunciated in the UN General Assembly Resolution No. 2028 (XX). Chagla underlined that 'a satisfactory agreement on non-proliferation of nuclear weapons would have to take into account the peculiar circumstances in which certain countries are placed' and 'India had a special problem of security against nuclear attack or nuclear blackmail'. He emphasised that this aspect must be taken into consideration before India's final attitude to the non-proliferation treaty was determined. India's vulnerable position on her northern frontier, and her susceptibility to nuclear blackmail from China and the USA are important factors governing the firm stand taken by India on the question of non-proliferation (Ibid: 44).

From outside, one may get the impression that Indira Gandhi's decision for nuclear test (May 1974) was deliberately timed before the 1975 NPT Review Conference, probably to remind the promoters of the NPT that India has apparently lost interest in arms control, and the test was a commitment against the NPT. The test may be viewed as a 'climax of India's opposition to the NPT'; it was simply the exercising of the verbal position it had taken during the treaty negotiation. The Congress Party stood by Indira Gandhi's all nuclear decisions. In the AICC Resolution on 6–7 December 1980, the Party expressed:

> [I]n safeguarding India's sovereign right to develop and use nuclear technology for peaceful purposes, India has continued to oppose discriminatory arrangements such as the NPT while urging the conclusion of a general convention of the total prohibition of the use or threat of use of nuclear weapons.
>
> (AICC Resolution, 1980)

An appraisal

When facing the difficult question regarding who should be given credit, and how much for India's nuclear weapons journey, it would be fallacious, and extremely difficult, to assign full credit to any single variable that was involved in the culmination of India's nuclear weapon status. Many political leaders, scientists and strategic thinkers have played their parts and deserve credit accordingly. To Surya Prakash, India's nuclear 'bomb has many fathers' (*The Pioneer*, 25 May 1998). If the Congress Party leaders conceived and nurtured it, the left party leaders questioned it; and at the end, the BJP leaders delivered it. Therefore, it would be irrational to give only the obstetrician more credit than is due. Rightly, 'all were party to the nuclear gatecrash' (Ibid.). Generally, most of the credit is bestowed on the Congress and BJP leaders and the leftist leaders are criticised being anti-nuclear.

Political leaders have played leading roles, but their preferences were filtered through the screen of the party they belonged to. However, India's nuclear policy cannot be explained without recognising the central role of many individuals (Perkovich, 2002:20). Since Nehru's time, Indian policy-making has relied on personality and other unpredictable factors. Though considerable strategic thinking lay behind every nuclear policy decision, all Indian political leaders, especially those who have tried to lay their hand on the nuclear option, seem to be aware of the Napoleonic strategic policy-making philosophy: every policy decision is 'two-thirds analysis, and one-third leap in the dark', but they tried their best to ensure that any leap made by them should not land in the dark. Probably that is the reason why India's nuclear weapons trajectory took a longer road and wider spectrum. In this protracted journey, Indira Gandhi's contribution is undoubtedly significant.

REFERENCES

- Bhatia, Shyam, *India's Nuclear Bomb*, Ghaziabad: Vikas Publishing House Pvt. Ltd, 1979.
- Chakma, Bhumitra, '*Toward Pokhran II: Explaining India's Nuclearisation Process*', Modern Asian Studies, CUP 39:1:189–236, 2005.
- Chengappa, Raj, *Weapons of Peace: The Secret Story of India's Quest to be a Nuclear Power*, New Delhi: Harper Collins, 2000.
- Das, Debak, 'Leaked Cables: Allende, Kissinger, Moynihan, and the Indian Nuclear Bomb', March 04, URL: https://www.wilsoncenter.org/blog-post/leaked-cables-allende-kissinger-moynihan-and-the-indian-nuclear-bomb, 2019.
- *Delhi Domestic News Service*, May 13, 1982, in FBIS-SA, May 14, 1982
- Dixit, J.N., *India-Pakistan in War and Peace*, New Delhi: Books Today, 2002.
- *Foreign Affairs Records*, Vol. 19, Nos. 9-10-11, Sep-Oct 1970, New Delhi, 1970.
- Frankel, Francine R. *India's Political Economy, 1947-77: The Gradual Revolution*, Princeton, New Jersey: Princeton University Press, 1978.
- Hoyt, Timothy D. 'Shades of Realism: The Impact of Personality, Ideology, and Systemic Pressures on Indian

Foreign Policy', in UK Heo and Shale A. Horowitz (eds.) *Conflict in Asia: Korea, China-Taiwan and India-Pakistan*, London: Praeger, p. 139–59, 2003.

- Jayakar, Pupul, *Indira Gandhi: A Biography*, Viking, Penguin India: New Delhi, 1992.
- Kapur, Ashok, *India's Nuclear Option: Atomic Diplomacy and Decision Making*, New York: Praeger Publishers, 1976.
- Karnad, Bharat, *Nuclear Weapons and Indian Security: The Realist Foundations of Strategy*, New Delhi: Macmillan India Limited, 2002.
- Kaul, T.N. *Ambassadors Need Not Lie: Some Aspects of India's Foreign Policy*, Vol. 1, New Delhi: Lancer International, 1988.
- Khilnani, Niranjan M. *Realities of Indian Foreign Policy*, New Delhi: ABC Publishing House, 1984.
- *Lok Sabha Debate*, 5th Series, Vol. II, No.5, March 17, cols. 130–31, 1972.
- Malcolmson, Robert, W. *Nuclear Fallacies: How We Have Been Misguided Since Hiroshima*, McGill-Queen's University Press, 1985.
- Malhotra, Inder, *Indira Gandhi: A Personal and Political Biography*, London: Hodder and Stoughton, 1990.
- McLean, Scilla (ed.), *How Nuclear Weapons Decisions Are Made*, London: Macmillan Press, 1986.
- *Minutes of the AICC Meeting*, December 6–7, 1980.
- Mirchandani, G.G. *India's Nuclear Dilemma*, New Delhi: Popular Book Service, 1968.
- '*Mrs Gandhi Reiterates India's Policy of Using Nuclear Power for Peaceful Purposes Only*', *India News*, August 1, p.1, 1983.

- Nanda, Krish, 'Will India Go Nuclear?', *Bulletin of the Atomic Scientists*, December, p. 39–41, 1971.
- NTI, (n.d.), 'Nuclear Chronology', URL: http://www.nti.org/e_research/profiles/India/Nuclear/2296_2803.html.
- Oberdorfer, Don, 'Pakistan Concerned about Attack on Atomic Plants', *Washington Post*, October 12, 1984.
- Perkovich, George, *India's Nuclear Bomb: The Impact on Global Proliferation*, New Delhi: Oxford University Press, 2002.
- Praful Bidwai and Achin Vanaik, *South Asia on a Short Fuse: Nuclear Politics and the Future of Global Disarmament*, Oxford University Press, 1999.
- Prakash, Surya, 'All Were Party to the Nuclear Gatecrash', *The Pioneer*, May 25, 1998.
- Ramanna, Raja, *Years of Pilgrimage: An Autobiography*, New Delhi: Viking, 1991.
- Sreedhar, (ed.), '*A.Q. Khan Interview with Nawa-e-Waqt, February 10, 1984*', in *Dr A.Q. Khan on Pakistan Bomb*, (ed.) New Delhi: ABC, 1987.
- Srivastava, C.P. *Lal Bahadur Shastri: A Life of Truth in Politics*, OUP: New Delhi, 1995.
- Srivastava, Govind Narain, ed., *Non-Alignment and Nuclear Disarmament*, Indian Institute for Non-Aligned Studies, New Delhi, May, 1985.
- Subrahmanyam, K. 'Indian Nuclear Policy – 1964-98: A Personal Recollection', in Jasjit Singh ed. *Nuclear India*, Knowledge World, New Delhi, 1998.
- Subramaniam, C. '*India's Defence Strategy in the Next Decade*, Transcript of address delivered by Shri C. Subramaniam,

Defence Minister, to the Officer Students of the National Defence College, New Delhi, on 29 October 1979.

- Sundarji, K. *Blind Men of Hindoostan*, New Delhi: UBS Publishers' Distributors, 1993.
- Varma, Shanti Prasad, Iqbal Narain, Chandra Prakash Bhambhri, *Fourth General Election in India: Political Surveys,* Volume 1, Orient Longmans, 1968.
- Zaidi, A.M and S.G., eds., *The Encyclopaedia of the Indian National Congress*, Vol. 22, 1972-73, *At War with Poverty*, New Delhi: S Chand & Company Ltd, 1984.

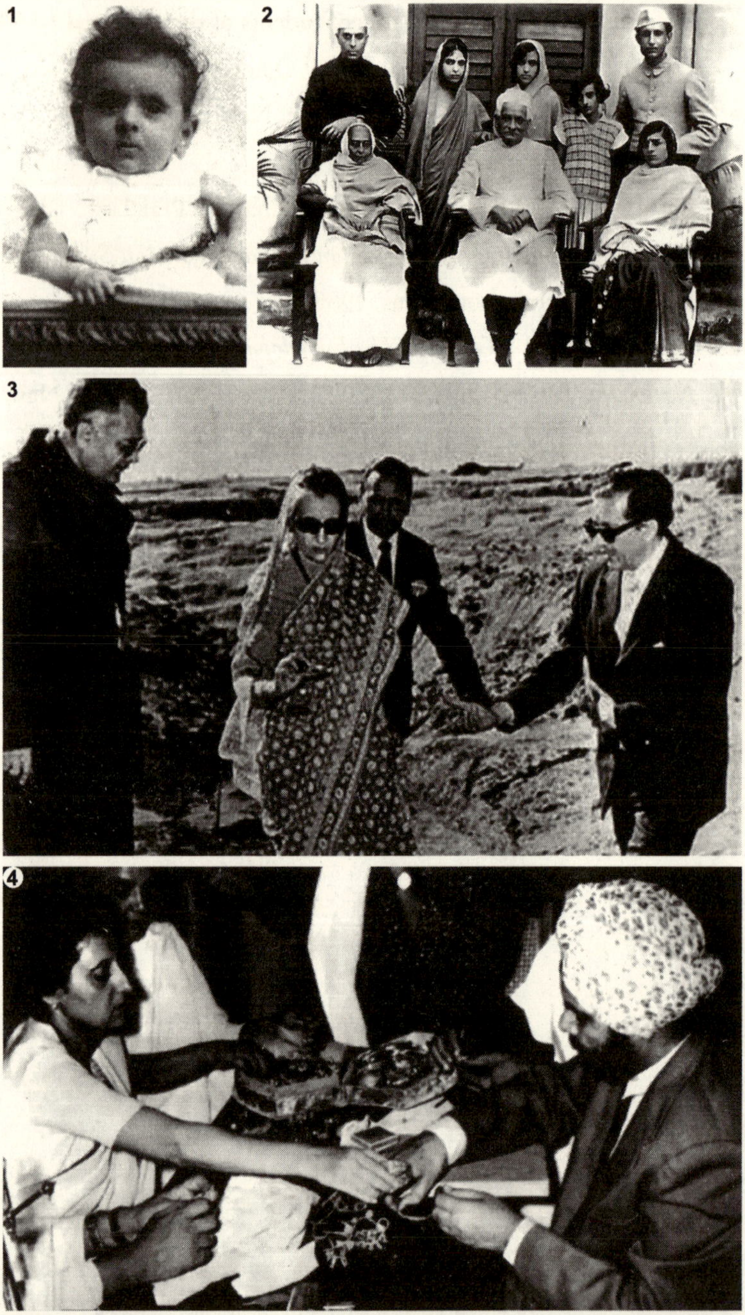

1. Childhood picture of Indira Gandhi 2. Nehru family 3. Indira Gandhi at the site of the test in Pokhran, 1974 4. Indira giving her jewellery in 1962 towards war efforts

ATAL BIHARI VAJPAYEE
The Ajatshatru Leader

Yasemin Develi

THERE IS A tendency in International Relations (IR) literature to ignore the impact personalities may have on foreign policy of a state. States are considered by traditional IR theories as the most prominent actors at the global stage. However, the reality is that decisions are taken by individuals, statements are written by human beings, information gathered, analysed, and passed on by individuals. It is after all a very human endeavour, so to speak. It is this fact that has driven many scholars since the second half of twentieth century, until now, to analyse personalities of leaders, group dynamics among decision makers, the role of bureaucracy and public opinion and so forth in order to understand, explain and predict state behaviour. Foreign policy, like domestic politics, is conducted by decision makers acting individually or in groups and therefore is grounded in humans (Hudson, 2005). Against this backdrop, we need to pay close attention to who makes the decision, as much as to what the decision is and the output/impact of that decision on other actors at the global level.

As heads of governments, Prime Ministers or Presidents play a central role in foreign policy decisions in accordance with their interest and capability in the field. The impact a leader can have on decisions depends on how much power and room they have to manoeuvre in complex constrained environments, their personal conviction to overcome obstacles, their overall response to constraints, their perception of other countries, and of their own as well. To understand any foreign policy decision, it is key to understand the very people behind any given decision, particularly the leader in the case of India.

Naturally, the role and impact of leaders in foreign policy decision making varies from one country to another for multitudes of reasons. However, in the case of India, the role and impact of leaders on foreign policy are amplified or rather more accentuated. This has several reasons. First of all, Indian foreign policy making has always been closely built around Prime Ministers. Starting from Jawaharlal Nehru, many Prime Ministers held the title of Foreign Minister during their tenures, and had their own small cliques of trustees to make important decisions. Indira Gandhi, for instance, is known for her 'kitchen cabinet' with which she discussed important domestic and external matters. Another reason why leaders matter in foreign policy making in India is the centrality of the Prime Minister's Office (PMO). It has always been the sole coordinating body linking various elements of foreign policy mechanism like a hub and spokes, to use an analogy. As the central figure in the PMO, the Prime Minister plays a decisive role in the process of this coordination exercise between different bodies. Thirdly, the general immunity of foreign policy matters from direct coalition scrutiny as well as public scrutiny frees the hands of decision makers and allows them to have a greater impact on foreign policy, though that has witnessed a change in more recent times. State governments and regional parties have been traditionally more interested in domestic politics than in foreign policy matters, with the exception of issues that concern India's bordering nations. Tamil Nadu's interest in certain issues related to Sri Lanka is an obvious example. Of course, the indifference of regional parties to external policies, gives the governing party an unparalleled wiggle room to make foreign policy decisions.

Here, India's nuclear tests spring to mind perhaps as one of the best examples that epitomises India's modus operandi when it comes to the foreign policy decision-making process. The decision to conduct the Pokhran-II tests was taken in secret, and preparations were made without anybody in the larger cabinet, let alone the coalition partners, knowing what was about to transpire. No doubt, this had a specific reason, but it still shows the power of the leadership in taking important decisions. This book talks about what happened before and after that monumental decision, and in this chapter, I will delve into the persona of the man behind that decision – Atal Bihari Vajpayee.

An interesting question to start with may be, was AB Vajpayee the right man in the wrong party or the wrong man in the right party? While such a question may evoke a negative connotation or an ambiguous feeling, it clearly gives one the impression that Vajpayee was not accepted by some quarters, depending on one's political or ideological views and leanings. However, in reality, he was one of the very few politicians who transcended party politics and almost magically gained people's unconditional love and respect.

Vajpayee's vivacious persona allowed him to outgrow his party and traditional political base, reaching out and appealing to a larger political spectrum. Vajpayee was a well-respected politician, and this made him the perfect candidate in his party for elections in 1996[4], 1998, 1999 and 2004. The fraction within his party and his complicated relationship with Rashtriya Swayamsevak Sangh (RSS) cadres and the Hindu nationalists at large did not come in the way of garnering their support. He represented the status quo, as well as change. He was part of the Delhi establishment, yet an outsider, a seasoned politician, a bold leader and an acceptable Prime Minister for a large coalition. By the time he left office, India was a nuclear power and one of the fastest growing economies in the world. The relations with the US were on the upswing. Vajpayee was an astute politician with formidable and exquisite credentials as former External Affairs Minister. He was interested, knowledgeable, and one may also confidently say highly versed in foreign policy and its intricacies which overshadowed his tenure as India's prime minister.

---

4 It was LK Advani who first publicly announced Vajpayee's name, much to everybody's surprise, in 1995, at the Shivaji Park meeting.

Childhood

It was on Christmas Day in 1924 that Atal Bihari Vajpayee was born, in Gwalior, Madhya Pradesh. His father, Krishna Bihari Vajpayee, was a poet and a school master, who interestingly enough, also became Vajpayee's roommate at DAV College in Kanpur as they were both studying law at the same time. Cooking for his unusual roommate during his student years turned him into a good cook. AB Vajpayee loved both cooking and eating, so much so that his sister, Kamala, with a big smile on her face, recounted a story from her fondest memories with him (Crux, 2009). She said once she caught him stealing freshly made *laddoos* (an Indian sweet) from the kitchen. Young Atal begged his sister not to tell their mother. Later in his life, his love for food, particularly Sindhi food, only grew manifold. He used to invite himself over to LK Advani's house for lunch or dinner to eat Sindhi *khichdi*[5] and Sindhi *kadhi*[6] (Crux, 2009). He himself once confessed that one of the things which he could not do while in office, but missed so much, was going to Old Delhi area to eat street food.

If food was his big passion, poetry was a bigger one. His poems touched the hearts of millions. One of his early poems was sung in some of the RSS shakhas regularly:

> *Hindu tan man, Hindu jeevan, rag, rag mera Hindu parichay.*
>
> (I am Hindu in body and spirit, my life is Hindu, Hindu is my only identity).

5 Indian dish made with rice and moong lentils.
6 A vegetarian yogurt-based curry famous throughout India and Pakistan.

In another poem, titled, 'We Shall Not Allow War', he assures Pakistan:

> Russian bombs or American, the blood spilt is the same.
> We have suffered, we will spare our children this fate.
> Never again will the sky rain fire.
> Never again will Nagasaki burn.
> We shall not allow war!

His poems were most times woven into his mesmerising speeches. He has said on many occasions that if he was not in politics, he would be a poet. As Prime Minister, he regretted having very little time for his biggest passion, poetry. Yet he wrote myriad poems throughout his life. He was witty, a great orator, seemingly a quality which ran in the family; his father too, liked playing with words. Vajpayee had three sisters – Kamala, Vimala, and Urmila – and three brothers, Prem, Suda, and Avadh. It was Prem with whom he was put in jail for a brief period of time during the Quit India Movement in 1942. Like many of his generation, his formative years coincided with the freedom struggle. His choice of subjects at school, his involvement in the Quit India Movement, his Arya Samaj and RSS affiliations at an early age, all show us a young man who got himself inextricably entangled in politics and was destined for something big.

His school years were spent in his hometown at Gwalior. Vajpayee studied at Saraswati Shishu Mandir and later went to Victoria College to get a BA degree in English, Hindi, and Sanskrit. His postgraduate education (MA) was in Political

Science from DAV College in Kanpur. As a student, Vajpayee liked reading classics, had a good command over English, though his special interest was Hindi, his mother tongue. Mastery of Hindi was to open many doors for him later in life, first in journalism and then in politics.

It is hard to separate Vajpayee's private life from his political life as he spent a major part of it in politics holding a wide range of positions. He was a journalist, a poet, an activist, a freedom fighter, a *pracharak* (full-time member of RSS), a founder member of three parties, a leader for various delegations – both national and international, an opposition leader, Foreign Minister and thrice Prime Minister of India. Throughout his eventful and active life, he made great contributions to India's political scene. He shaped Indian politics, and in turn was shaped by the social and political disturbances of his day.

Political life

Vajpayee's political life started at an early age when he joined Arya Kumar Sabha, the youth wing of Arya Samaj. He subsequently became its General Secretary in 1944, just three years before the partition and independence of India. He was particularly influenced by Arya Samaj philosophy. Later in his journey, he met Babasaheb Apte, one of the early pracharaks of the RSS and joined the organisation in 1939 at the age of fifteen. As a young boy, along with his brother Prem, he used to attend the *shakhas* (branches of RSS). He became a very disciplined *swayamsevak* (volunteer) in the organisation and his commitment and hard work made him a pracharak in the same year India gained independence. Before the RSS, he was, for a brief period, with All India Students Federation affiliated to the Communist Party of India (CPI). Like many of his generation, Vajpayee was also influenced by Marxist ideology, though this did not last long. Then he became an RSS member. During this time, he was more under the influence of Hindu nationalism, political worldview of Hinduism, without however taking to any extremist view.

His exceptional command of Hindi and his relationship with the RSS got him a job as an editor at several RSS journals which belonged to Deendayal Upadhyaya. After leaving law school, due to the political turmoil caused by partition, he worked as an editor in several journals, *Rashtra Dharma, Panchjanya, Swadesh* and *Veer Arjun*.

The assassination of Mahatma Gandhi and the subsequent ban on RSS compelled the organisation to acquire political power for its survival (Andersen and Damle, 2018), a move that was contrary to RSS's early doctrines of remaining a cultural organisation, which in fact considered politics a dirty world. Nonetheless, having been forced to go underground after the ban, and not being happy with the Congress policies, some quarters of the RSS volunteers led by Syama Prasad Mukherjee decided to have a foothold in politics and formed the Bharatiya Jana Sangh (BJS) in 1951. It was the political arm of the Sangh Parivar that sought to provide a platform to convey its message and to implement its national agenda. This young nationalist political party was founded on the premise of offering an alternative to the Congress.

Initially Vajpayee was not involved in the party in an official capacity; he was seconded by the RSS to provide a helping hand to the young party. His skillset, oratory and personality caught the party president Mukherjee's attention. Mukherjee made Vajpayee his aide to help with his party work. In those years, Mukherjee was fiercely opposing the status of Jammu and Kashmir under the Liaquat-Nehru pact. He went to Kashmir to protest against the permit system and Vajpayee accompanied him on this journey. Mukherjee's arrest and subsequent death left an indelible mark on Vajpayee and his general persona.[7] It was only after Mukherjee's death in detention that Vajpayee became fully and officially involved in the party's work. Soon after, he stood for the 1957 elections, contested and won the Balrampur seat. This was to be the beginning of a remarkably long and successful political career. Vajpayee was elected to the Lok Sabha ten times – in 1957, 1967, 1971, 1977, 1980, 1991, 1996, 1998, 1999, and 2004, and twice to the Rajya Sabha, first in 1962 and second time in 1986.

Throughout his extraordinary political life, he commanded respect from political rivals and allies alike for his liberal worldview and commitment to democratic ideals. Upon entering parliament, Vajpayee took a particular interest in foreign policy and fiercely criticised Nehru on many occasions.

7 He penned down this story as such: 'As the policemen led him away, he turned towards me and said, "Go back Vajpayee. Go back and tell the people that I have entered Jammu and Kashmir State, though as a prisoner." That was the last time I saw this great nationalist, but his words still ring in my ears, spurring me on in those moments when despair, tiredness and fatigue threaten to overwhelm me.'

He was very active in Parliament and his innate and effortlessly flowing oratory put him in pole position, as the best person to articulate the party's grand ideals, policies, agenda and manifesto.

He rose to the top in his own party while showcasing his political acumen and oratory skills on the floor of the Lok Sabha in those days. After the mysterious death of the party president Upadhyaya, Vajpayee became the president of BJS in 1968. The subsequent years were eventful – India–Pakistan war in 1971, India's first 'peaceful nuclear tests' in 1974 and then the emergency rule imposed by Indira Gandhi in 1975. Vajpayee was detained along with his party colleagues and other opposition leaders. His health deteriorated while in detention. Due to this, he was transferred to a hospital in Delhi from his prison in Bangalore. By the time Indira Gandhi lifted the emergency restrictions in 1977, BJS decided to merge with other opposition parties to defeat the Congress in the upcoming elections. Vajpayee and his colleagues contested the elections under the Janata Party coalition and defeated the Congress.

A new government was formed and Morarji Desai became the first non-Congress Prime Minister of India. AB Vajpayee was appointed as the External Affairs Minister, a position he held for two years and relinquished because of the 'dual membership' controversy, a condition forced by secular members of Janata Party on Vajpayee, Advani and other former BJS members so as to make them choose between staying in power and being a member of the RSS. He, along with Advani and some erstwhile BJS members, left the Janata Party and formed a new party in 1980 – the Bharatiya Janata Party (BJP). For the first time, Vajpayee made 'Gandhian socialism' the cornerstone of this new party.

As External Affairs Minister (EAM), Vajpayee engaged with Pakistan and China more closely and became the first Indian EAM to visit China since the 1962 war. However, the biggest impact he had was on relations with Pakistan. Purno Sangma once shared in Parliament his experience in Pakistan, when he went to attend a conference on People-to-People Contact:

> I am happy to tell you here that one of the gentlemen in Pakistan asked: 'Mr. Sangma, do you know who is the most popular politician of India in Pakistan?' I asked him: 'How do I know?' He said: 'The most popular politician of India in Pakistan [is] Shri Atal Bihari Vajpayee.'
>
> (Twelfth Lok Sabha Discussion, Session 1, 29 May 1998)

Sangma went on to explain,

> They said that today you are having a Conference on people-to-people contact, but this process of people-to-people contact between India and Pakistan was started by Shri Vajpayee when he was the Foreign Minister of India. He went to Pakistan. He liberalised and streamlined the visa procedure. He made the procedure for getting visa easy so that people from Pakistan could come to India and people from India could go to Pakistan. That was the initiative taken by Shri Atal Bihari Vajpayee. That is why he was called the most popular politician in Pakistan.
>
> (Ibid.)

The BJP's first ever elections in 1984, after leaving the Janata Party experiment behind, were, by any standards, an absolute disaster and a bitter disappointment, as they managed to win only two seats in the Lok Sabha. This political underachievement was largely attributed to the 'sympathy wave' which Congress harnessed after Indira Gandhi's assassination by her Sikh bodyguards. Just over a decade later, the party won the 1996 elections and Vajpayee became Prime Minister for the first time. However, his premiership was short-lived as it only lasted thirteen days. The numbers were not in his favour to form a government. His second term as Prime Minister started in 1998. One of the decisions he took soon after assuming power was to give clearance for the preparations of the nuclear tests.

Leadership

> The question is not of 'right' and 'left'; the issue really is Democracy versus Totalitarianism. The choice is not between 'progressivism' and 'reaction'; the people have to choose between forces uncompromisingly pledged in their loyalty to the nation with extra-territorial loyalties. Differences are bound to remain in the country, but the Indian nation cannot afford to be divided in its basic commitment to Nationalism and Democracy.
>
> (AB Vajpayee, 16th All India Session, Chanakya Nagar, Patna, 28–30 December 1969)

'He was a person who could bring out the best in others,'

said Arun Shourie, aptly describing Vajpayee who invariably allowed everybody to speak their mind freely and was more of a listener than a lecturer (NDTV interview, August 2018). He was a man of few words with distinctively long pauses. He had a good sense of humour. One could oppose his views in private or in meetings without any hesitation. He was a leader who was vehement, yet tolerant of opposing views, an *ajatshatru* (one without an enemy). He encouraged professionalism and expertise and surrounded himself with strong personalities like Brajesh Mishra, who was an astute diplomat; like Advani, who was a great organiser and a hard-line RSS ideologue; like Jaswant Singh, whose command over English was equivalent to that of Vajpayee's Hindi; like Yashwant Sinha, a master of details, and many such others. These personalities were instrumental in his success in managing domestic and external policies.

Vajpayee gave enough room to everybody to function freely and contribute to the best of their ability. There was very little interference on a day-to-day basis, either directly from Vajpayee or from his PMO. He did not feel threatened or intimidated by the constraints he encountered. He was, in a sense, a smart player, a consensus seeker. This is not surprising as he was one of the most experienced politicians in India at the time. Once when questioned about his biggest principle in politics, Vajpayee replied, 'being genuine is my principle' (NDTV, 16 August 2018). Certainly, Vajpayee came across as an embodiment of everything genuine and authentic. He believed in civil discourse, in the culture of civilised debate, opposition not enmity, discussion not fight.

Despite winning the 1996 elections as the single largest party, Vajpayee failed to cobble up a coalition as the other parties did not want to affiliate themselves with a 'right wing' party like the BJP. Therefore, he had to resign before the confidence vote. The lessons learned from this experience enabled him to bring on board twenty-three allies and form the National Democratic Alliance government in 1998. This was only possible thanks to the acceptability of Vajpayee as a moderate leader within a right-wing party. Through his mastery of compromise, gentle cajoling and persuasion, he proved to the people of India that coalition governments could work.

Vajpayee's liberal approach to politics, his moderation, and his affection for Nehru made many believe that he was 'Nehru of the right'. He certainly shared some ideals with Nehru. That is why many people who were opposed to what the BJP stood for, tried to accommodate Vajpayee owing to his 'Nehruvian' style. Definitely, his affection for Nehru, or more so his moderation, aroused some resentment within the hardliners in his party and in the RSS.[8] However, decades spent in Parliament, countless election campaigns, having to form coalitions with diverse political parties meant compromise for him. He did admire Nehru's personal qualities as a leader, yet opposed him politically. Vajpayee belonged to the RSS, and he repeated his position on many occasions. He once said in Parliament, 'Time and again it has been said that Vajpayee is good, but the party is not.

8 He was dubbed as mukhauta (mask) of the BJP by party ideologue Govindacharya who later claimed he was misunderstood.

That is all fine, but what do they intend to do about "the good Vajpayee"?' (NDTV, 18 August 2018).

Atal Bihari Vajpayee's personality allowed him, and in fact impelled him, to walk the delicate ideological tightropes successfully. He did disagree with Nehru on domestic and external matters, yet he admired the man's personality, his leadership, which he often fondly spoke of. One can infer from Vajpayee's memories of Nehru an implicit gratitude and an explicit reverence. He recounted:

> I took a personal interest in foreign affairs and here I must note that Nehru, despite all his shortcomings, was a democrat at heart. Once I spoke at length on the floor of the House, castigating the government's foreign policy. I spoke in Hindi from the Opposition back benches. When it was the Prime Minister's turn to reply to the debate, I was pleasantly surprised to note that he had listened attentively to all the points raised by me. In fact, although he spoke in English, he replied to my points in Hindi. A second point I would like to make is Nehru's large-heartedness. Although we berated the government day after day and did not have the numerical strength to make a difference, Nehru made it a point to introduce me to foreign visitors at official dinners.
>
> (*Frontline*, 9 August 1997)

Vajpayee maintained this culture of tolerance as much as he could while in office. He understood that in a country as diverse as India, there would be difference of opinion, but all should serve the country's interests. In his famous speech in

Parliament in 1996, he gave a glimpse of his idea of politics. He talked about the time when Narasimha Rao sent him to Geneva as the head of a delegation. He was the opposition leader at that time. He said, '[Pakistanis] thought I would not support the government. They think opposition is obsessed with toppling the government. That is not part of our tradition, nor our nature. I want that kind of tradition to continue.' And this tradition was present in his modus operandi throughout his political life, both in opposition and in power.

He repeatedly explained his ideological position. But in order for the BJP to reach out to a wider political base, acquire new allies and stay in power, it had to tone down its zealot agenda, abandon some of its pledges like abrogation of Article 370, uniform civil code, building a temple in Ayodhya, etcetera. Vajpayee was the right man for the job as he believed in the merits of compromise. Vajpayee's strategy paid off and the NDA coalition successfully formed the Government in 1998. However, one thing that stayed on the common agenda, and on which the BJP could not and would not compromise, was the nuclear policy. The party, time and again, pledged to re-evaluate India's nuclear policy (Election manifesto, 1991, 1996, 1998).

Nuclear Decision

'Action is a process to reach a goal; action may reflect tumult but when measured and focused, will yield its objective of stability and peace (Lok Sabha Debate, session 2, 27 May 1998).' These were the concluding lines of Vajpayee's speech in Parliament after the nuclear tests, a quote he took from the Gita.

Before the 1998 elections, the BJP, along with its alliance partners, published a common national agenda. In this document, the BJP's compromise on many core issues like Ram temple caught the media's attention. Above all, it was the nuclear policy that made the headlines. The BJP and its allies pledged to re-evaluate the nuclear policy and 'induct nuclear weapons' if they won the elections, which caused a concern and reverberated in the region and across the world. Remarks and reactions from Pakistan and the US were also making the headlines in those days. For decades, the BJP criticised previous governments for endangering India's national security by succumbing to pressures from the West and by not exercising the nuclear option. The party firmly believed that nuclear weapons would deter China and Pakistan from attacking India. Vajpayee, however, described the security environment with these words:

> The end of the Cold War marks a watershed in the history of the twentieth century. While it has transformed the political landscape of Europe, it has done little to address India's security concerns. The relative order that was arrived at in Europe was not replicated in other parts of the globe.
>
> (Lok Sabha Debate, session 2, 27 May 1998)

Furthermore, famously promising no compromise on national security, Vajpayee had four rational but difficult options on the table regarding the nuclear tests when he came to power. First, signing the Non-Proliferation Treaty (NPT) of 1970; second, maintaining nuclear ambiguity of previous governments; third, declaring India as a nuclear state without testing; and fourth, conducting nuclear tests which would catapult India into the select club of nuclear powers and allow Vajpayee to formally declare India as a nuclear state. While the first option would be relatively easy to exercise, it was not acceptable to India, whereas maintaining 'ambiguity' would not be an option if India did not test quickly, either. Because of the indefinite extension of the NPT in 1996, there would have been a nuclear club from which India would be excluded forever. Thus, it was not a viable option in the long run. The third option of declaring India's nuclear status without testing, however, would put India in a precarious position, jeopardise its security and damage its standing in the world. On the other hand, testing nuclear weapons would be an even riskier option on a multitude of levels. Therefore, Vajpayee's India was inextricably stuck between a rock and a hard place. For the leader of a coalition government, some of these options would have perhaps seemed easier than conducting controversial tests and declaring his country as a nuclear state.

For decades, the BJP had included in its manifestos the option of conducting nuclear tests and declaring India a nuclear power. It is widely shared that Vajpayee tried to initiate such tests as soon as he came to power in 1996, but unfortunately for him, because his tenure lasted just thirteen days, forcing him to resign from office, he could not carry out his desire to see India conduct nuclear tests for a second time in the nation's history. Of course, this was not the first time for an Indian

Prime Minister to wish to press ahead with conducting the tests. It is common knowledge now that previous governments tried, failed, and/or put off conducting nuclear tests for several reasons, chief among them the fear of being caught by hovering US spy satellites that were able to detect suspicious military activities anywhere in the world, and the fear of opening themselves up to heavy US sanctions.

Though it was during his tenure that India finally conducted the tests, Vajpayee never tried to claim all the credit for himself, rather, he magnanimously acknowledged the contributions of preceding governments in keeping the nuclear option open. In his statement after the tests, he remarked, '[in] 1974, we demonstrated our nuclear capability. Successive governments thereafter have taken all necessary steps in keeping with that resolve and national will to safeguard India's nuclear option (Lok Sabha, session 1, 1998).' Vajpayee's acknowledgement is generous, heartfelt and matter-of-fact. It was after all Nehru who laid the foundation for the nuclear programme, and Nehru's successor, LB Shastri, who gave the orders to press ahead with a nuclear test, but passed away soon after. Indira Gandhi finally successfully managed to conduct the tests but fell short of declaring India a nuclear power, and instead declared the tests during Operation Smiling Buddha as 'peaceful explosions'. On the other hand, Rajiv Gandhi, the sixth Prime Minister of India, campaigned for global disarmament. During his state visit to the US in 1987 (Reagan Library, 2017), standing by President Reagan, he stated, 'We do not have nuclear weapons, we do not want nuclear weapons and we certainly do not want nuclear weapons in our neighbourhood. On our part, let me assure you Mr President, and the people of your country that we have no intention of producing nuclear weapons unless constrained to do so.' Yet, he strongly supported the program and kept the nuclear option open.

Successive governments also followed suit. As Prime Minister, Narasimha Rao gave his consent for conducting nuclear tests, but the Central Investigation Agency (CIA) detected the preparations and activities early on, which led to the US warning the Indian government. Rao acquiesced to the mounting pressure. After all, it is believed he needed the US for the liberalization policies he had envisioned for the Indian economy. Although this attempt failed, Rao warned Vajpayee about the risks of being caught by US spy satellites. It can be said that, as Prime Ministers, even Deve Gowda and IK Gujral had retained the option of testing, albeit tacitly, by not signing the CTBT. With all these experiences and more, Vajpayee knew that such a risky venture could only be successful if top secrecy was maintained at all times and in every step leading up to the day of execution. Learning from previous failed and last-minute aborted attempts, Vajpayee knew too well that only a select few people had to be privy to his decision before the test, some of whom were informed by Vajpayee personally just days before the test (Singh, 2006).

After the nuclear test, the Vajpayee government paid particularly careful and considerate attention to getting its message to the world. The media and public relations (PR) strategy of the time was to explain that India's tests and nuclear weapons were not directed against any specific country. Trying to justify Vajpayee's decision and allay US concerns and fears at once, Principal Secretary to the Prime Minister, Brajesh Mishra said at the time, 'I hope the US will understand that the tests were meant for security of the people of India' (Shankar, 1998). This effort to keep the US in the loop, and more importantly ease the already imposed sanctions and build a better relationship with the superpower, gained further momentum. The famous Talbot-Singh talks were part of a larger strategic plan to engage more closely with the US, and to do so, Manmohan Singh first needed to explain India's reasons for making the decision to conduct the nuclear tests. Such intense, high-level diplomacy and the many carefully calibrated efforts by Vajpayee and his team culminated in the 2005 US–India nuclear deal during Manmohan Singh's tenure as Prime Minister.

Pakistan followed India's suit in 1998, just weeks after India's successful tests, and conducted its nuclear test, declaring itself as a nuclear power. Though this created a new security concern in the region, Vajpayee took several bold initiatives to improve relations with Pakistan. He first inaugurated the Delhi–Lahore Bus Route in February 1999. However, this atmosphere of 'peace' soon vanished because of the insurgency in Kargil which led to a full-blown conflict in May 1999. These events did not prevent Vajpayee from taking further steps to bring peace to the region either. After all, one could 'choose friends, but not neighbours' as Vajpayee once said, and India would have to learn to live with them, so to speak. His second effort to resolve outstanding and thorny issues with

Pakistan was the 2001 Agra Summit which again failed to yield fruit. Relations with the US, on the other hand, started on the wrong foot due to India's surprise nuclear tests, but soon started changing. The real shift from 'estranged democracies' to 'engaged democracies' happened during Vajpayee's tenure.

Vajpayee's time in office ended with a surprise defeat in the 2004 elections. Soon after the result, he announced his retirement from politics. His health deteriorated, and in 2009, he suffered a stroke, which sadly impaired his speech. Vajpayee, a towering figure in Indian politics, died in August 2018, leaving an indelible mark on India.

Conclusion

Conducting nuclear tests and calling India a nuclear power is a controversial subject even today. While some praise the fact that India is among the nuclear haves, others claim it created more tension in the region, and gave Pakistan the legitimacy to go ahead with their tests. Nevertheless, India is in the nuclear club. Many prime ministers contributed to India's nuclear programme, but Vajpayee is undeniably the key person in 'the decision'. His contribution to India's global outlook is tremendous, and he will always be remembered for that.

REFERENCES

- Andersen, Walter K. and Damle, S. D. *The RSS: A View to the Inside*, Penguin Random House, 2018.
- Crux. Atal Bihari Vajpayee: A True Statesman of Modern India | Rare Interviews | Crux Files, 24 December 2019.
- Election Manifesto, BJP, 1991, 1996, 1998.
- Hudson, Valerie, 'Foreign Policy Analysis: Actor Specific Theory and the Ground of International Relations', *Foreign Policy Analysis*, Vol.1(1): 1–30, 2005.
- Kux, Dennis, *India and the United States: Estranged Democracies, 1941-1991*, National Defense University Press, 1992.
- Lok Sabha Discussion, Session 1, 29.05.1998, URL: http://loksabhaph.nic.in/Debates/Result12.aspx?dbsl=157
- NDTV, Atal Bihari Vajpayee Wasn't Jealous, Or Insecure, Says Arun Shourie, 16 August, URL: https://www.youtube.com/watch?v=iHU6SSXDvPQ, 2018.
- NDTV, Never Thought I Would Be A Politician, Always Wanted To Be A Poet: Vajpayee, Aired: April 2004, URL: https://www.youtube.com/watch?v=DuRkJS7M7F0, 2018.
- NDTV, Truth vs Hype: The Vajpayee Paradox, 18 August, URL: https://www.youtube.com/watch?v=_YpIncvfqkg&t=5s, 2018.
- Quint, URL: https://www.youtube.com/watch?v=Bn9DhR7pYsQ&t=777s. URL: http://ataljee.org/quotes.html, 2012.
- Reagan Library, President Reagan's Remarks after Discussions With Prime Minister Rajiv Gandhi on October 20, 1987,

URL: https://www.youtube.com/watch?v=0nhbe5Sv4hs, 2017.

- Singh, Jaswant, 'One Morning in May' in *A Call to Honour*, Rupa Publications India Pvt. Ltd.: New Delhi, 2006.
- Shankar, Jay, 'Indian government, industry urge US not to impose sanctions', Agence France-Presse, 11 May, 1998.
- Vajpayee, A.B. 'The BJP's onward march', *Frontline*, August 9, URL: https://frontline.thehindu.com/other/article30158635.ece, 1997.
- Vajpayee A.B. 'The Evolution of India's Nuclear Policy', speech at Lok Sabha debate 27.05.1998, session 2, URL: http://loksabhaph.nic.in/Debates/Result12.aspx?dbsl=249, 1998.

1. Atal Bihari Vajpayee and his family 2. Vajpayee with Narasimha Rao
3. Vajpayee at Pokhran nuclear site (HT, 1998)

SECTION C

THINKERS STRATEGISTS

K SUBRAHMANYAM
The Scholarly Strategist

Soumya Awasthi

KRISHNASWAMY SUBRAHMANYAM WAS a key figure in international strategic affairs and a strong voice in national security matters. KS wore many hats throughout his life, first as an Indian civil servant, then as a strategic analyst and finally as a journalist in his later years. He was considered an advocate of Realpolitik, earning him the title 'India's Kissinger' (Foreign Policy). He was one of the early advocates of developing a bomb in the era of Gandhi and Nehru. Even though *Subbu*, as his friends called him, advocated that India must possess a bomb, he also understood very well the destructive power of the same. Therefore, he was one of the two proponents of the 'No-First-Use' policy, the other being General Krishnaswamy Sundararajan, commonly known as General Sundarji, Chief of the Army Staff of the Indian Army from 1986 to 1988 (SSRN).

Subrahmanyam was born on 19 January 1929. He grew up in Tiruchirappalli in a conservative Tamil family. His father was a principled and disciplined man who was the headmaster of a Tamil medium school. Subrahmanyam's father got him admission to a Tamil medium school, though he was interested in learning English. The eldest of six siblings, he did not have a very comfortable life. He used to walk almost three to four kilometres barefoot to school every day and managed with few clothes. These hardships made him understand the value of things in life.

He loved books and sitting in libraries; so to escape his strict father's control, he deliberately chose to study at the University of Madras and received an MSc degree in Chemistry from Presidency College in 1950. According to his daughter, Sudha Subrahmanyam, he got his first pair of slippers only when he joined college, as wearing slippers in the chemistry lab was mandatory.

He was ranked first in the Indian Civil Services Examination and was appointed to the Indian Administrative Service in 1951. He held many posts while in service, including the post of Secretary in the Ministry of Defence from 1962-65 and 1979-80. He was appointed as a Rockefeller Fellow in Strategic Studies at the London School of Economics in 1966. As a young visiting fellow from India, it is recalled that he stood up and spoke out on the 'hypocrisy' of the West in the presence of the then British Secretary of War (Brookings, 2018).

He went on to hold several senior positions in the Government of Tamil Nadu and the Government of India, including Chairman of India's Joint Intelligence Committee in New Delhi; Member of the Board of Revenue, Government of

Tamil Nadu; Home Secretary, Government of Tamil Nadu; Additional Secretary, Cabinet Secretariat, New Delhi.

When he was chairman of the Joint Intelligence Committee from 1977-79, he drafted the Standard Operating Procedure in case of a hijack. It is said that Subrahmanyam was on board an Indian Airlines flight (IC 421) on 24 August 1984 when the plane was hijacked; the arrested hijackers later claimed in court that it was Subrahmanyam who 'planned the entire hijacking to examine nuclear installations in Pakistan' (*Hindustan Times*, 2011).

He was appointed the Secretary for Defence Production in the Ministry of Defence. He then served as the Director of the newly created Institute for Defence Studies and Analyses (IDSA) in New Delhi from 1968-75 and later from 1980-87. While holding the position of Director, he contributed immensely through his articles, columns and books, greatly influencing Indian strategic thought.

Interestingly, *Subbu*, as he was popularly known amongst friends and colleagues, wanted to teach at the Jawaharlal Nehru University (JNU). However, he was not permitted because he did not have a doctorate. He recognised the irony in that when Cambridge, where he eventually received a professorship, had no such reservations regarding his academic credentials (*Outlook*, 2022). In 1987, he returned to England as a visiting professor and Nehru Fellow at St John's College in Cambridge. KS participated in several United Nations and other multilateral research groups between 1974 and 1986, covering topics including disarmament, nuclear deterrence, and Indian Ocean politics (JWSR). He was also a member of the

Pugwash Institute in India and regularly organised discussions on nuclear proliferation (*Asian Affairs*, 2000).

Subrahmanyam's intellectual contribution through various articles, books and speeches highlights the passion with which he defended what he thought to be in India's absolute national interest. KS wrote several books and articles throughout his career; *The Liberation War* (1972), *Nuclear Myth and Realities* (1980), *Superpower Rivalry in the Indian Ocean* (1989), and *India and the Nuclear Challenge* (1986) are some of his most famous works.

The American intellectual circle considered him India's 'no. 1 anti-American,' and some even labelled him a 'Communist'. This was because he was 'frequently on a collision route with Washington', especially after the United States suddenly cut off assistance to India during the Indo–Pak war of 1965. At this time, he was serving as the Deputy Secretary of the Ministry of Defence.

Subbu appeared stern and reserved but was very polite; he admired people like MK Gandhi and K Kamaraj. He always believed in the country's youth's potential; he encouraged young minds and was confident about the nation's future. In the recent past, despite his advanced age, his vibrant and optimistic vision of India's present and future juxtaposed fundamentally with his colleagues and predecessors, who were primarily conservative in their viewpoints.

Another interesting fact about Subrahmanyam is that he declined the honour of Padma Bhushan by the government of India in 1999, stating that bureaucrats and journalists should not accept government awards. He was one of the most influential Indian media figures and was featured in *India Today* magazine's 'High & Mighty' listing in 2006. In 2007, IDSA instituted an annual 'K Subrahmanyam Award' for contributions to Strategic Affairs as a tribute to the contributions made by K Subrahmanyam.

In one of her blogs, Vidya Subrahmaniam, who worked under Subrahmanyam during his stint in the *Times of India* in his later years, recalls how this 'distinguished-looking editorial consultant' had been introduced to their team 'as the father of the Indian bomb'. She recalls that initially, they found him intimidating and difficult to convince. But with time and constant interactions, it was found that he was full of humility; he treated each of his staff equally and had no interest in any personal advancements. Soon he was referred to as *Bomb Mama* (Bomb Uncle).

He was a disciplined man who considered punctuality the greatest virtue; he never missed his deadlines, and this despite being a heart patient and having severe diabetes. His wife, Sulochana Subrahmanyam, sharing an incident about Subbu's obstinacy, recounted how 'once, after a major heart surgery, he wrote and submitted an article for a publication from the hospital bed'. Later, even when he was diagnosed with cancer and was undergoing long chemotherapy sessions, he never stopped writing. He used to write in long hand and his writing became indecipherable, but he always remained positive saying, 'it was a challenge and a test of his endurance,' writes Vidya in her blog, *Farewell to Bomb Mama*.

K Subrahmanyam was a realist; he was a patient listener and never disagreed with anyone outright, but stood firm on his ground; he was never dismissive of counterviews, and replied with logical and rational reasoning. He believed that any nation, and specifically India, must involve itself in assessment and should adapt itself to the changes in the world order because he believed, 'if we leave the world alone, the world will leave us alone'. Therefore, a nation must look at the worldview and bring changes in policies. It is because of this pragmatic realism that, in 1971, he had unequivocally supported his country's alignment with the Soviet Union. But then, after the 2005 civil nuclear deal with the US, he emerged as one of the most fervent supporters of an accord with the United States.

Contribution to the nuclear policy of India

Subrahmanyam's role in India's nuclear program is another example of his pragmatic approach surpassing the dichotomy between idealism and realism. However, he always remained a nuclear peacemaker favouring only a limited program; he opined that nuclear weapons are a mere instrument to achieve India's more extensive national interests. In a famous interview with Amitav Ghosh in 1998, KS said, 'a nuclear weapon is like a million pound note, is of apparently no use – you can't stop small wars – but it gets you credit and that gives you power to intimidate' (*Outlook*, 1998).

Subrahmanyam was among the leading proponents of India's nuclear programme and use of the nuclear weapons option, which started with the country's first 'peaceful' nuclear test – the Smiling Buddha – in 1974 and culminated with the 1998 'Shakti' series of weapons tests. His views on India's emergence as a nuclear power with military and civilian capabilities impacted nuclear proliferation and nuclear disarmament. His strong opinions and incisive critique of the discriminatory nature of the Nuclear Non-Proliferation Treaty and Comprehensive Nuclear-Test-Ban Treaty regimes are well-known and have received much commentary and criticism.

He frequently criticised the five nuclear-armed governments for imposing the Nuclear Non-Proliferation Treaty on everyone else while rejecting any proposals for nuclear disarmament, notably the one made by India in 1986. He frequently used the phrase 'nuclear apartheid' to describe this position. He came up with the term 'nuclear ayatollahs' to describe the western non-proliferation lobby, which frequently accused India of having a nuclear programme and subsequently for agreeing to cooperate with the US on nuclear issues in 2007.

K Subrahmanyam was an active participant and one of the key contributors to India's nuclear policy. He was appointed as the Convener of India's first National Security Council Advisory Board (NSCAB) established by the Atal Bihari Vajpayee government in 1998. The board drafted India's draft nuclear doctrine, which governs all policy aspects regarding the use and deployment of India's nuclear arsenal. The declaration of the 'No-First-Use' policy and the consequent directive that India's nuclear weapons would primarily be oriented around a second-strike capability was its most significant aspect. Subrahmanyam had been arguing for India assuming a 'no-first-use' posture ever since the Shakti tests in 1998 and even earlier in 1974.

The thinkers and ministers across the country often used Subrahmanyam's arguments to support India's entry into the nuclear club, despite India's non-violent Gandhian origins and foreign policy roots in Nehruvian pacifism. This ingenious and paradoxical stand has been adversely commented upon, being labelled as 'moral exceptionalism' on non-proliferation by several commentators. Strangely enough, Subrahmanyam also argued that Pakistan too should look to develop a limited nuclear programme of its own – to establish its own deterrent against conflict. This was met with great suspicion in Pakistani circles and seen as an insult or a challenge. He also concluded in the 90s that India needed only 150-odd warheads to achieve minimum deterrence, a point acknowledged by many experts in the field. This was achieved by 1990 itself.

In 1979, as Chairman of India's Joint Intelligence Committee, Subrahmanyam voiced his opinion to restart India's nuclear weapons programme. The programme had been shut down in 1977 by Prime Minister Morarji Desai upon assuming office, mainly because he sought to return to India's pacifist foreign policy roots. Subrahmanyam's note was prompted by Indian Intelligence's latest estimates of progress in China and Pakistan's nuclear programmes. It was discussed and shot down in a Cabinet meeting, principally by Morarji Desai and Atal Bihari Vajpayee, then India's External Affairs Minister. Ironically, Vajpayee, immediately after taking over as the Prime Minister, authorised the 'Shakti' series of nuclear weapons tests in March 1998.

In section two of his compilation work, *Shedding Shibboleths: India's Evolving Strategic Outlook*, Subrahmanyam argues for a minimum deterrent policy and posture. Here, combining bits and pieces of the standard history of the development of nuclear weapons in the United States and the history of strategic deterrence during the Cold War, he builds his case in a roundabout way. The book includes case studies of countries like France, China, Israel, etcetera, coupled with riffs on the CTBT and the controversial 'No-First-Use' principle. Referring to the South African experience, which resonates with his views, Subrahmanyam says that a small nuclear force is more than adequate for deterrence purposes, 'because nuclear deterrence can be exercised at a very minimum level... need not result in a costly arms race... and [that] above all, uncertainty plays a crucial role in projecting deterrence'.

KS on the Theory of Nuclear Deterrence

Deterrence as a concept is as old as the ability of human beings to inflict pain on their fellow human beings and to anticipate the other person's capacity to inflict such pain. Pain need not be physical. It can also include destruction of and suffering to one's near and dear ones and what one cherishes the most.

According to K Subrahmanyam, deterrence should be understood in that we all, as human beings, have a stake in the survival of our respective national societies. He borrows from Robert Oppenheimer's analogy that the two superpowers are like two scorpions in a bottle – one cannot destroy the other without being destroyed in the process.

Since the beginning, nuclear deterrence and nuclear weapons have been projected as the currency of international power, and bring about a regime in which the use and threat of use of nuclear weapons by developed nations of the world will be legitimate in terms of conventions and laws of war, but not for others.

Nations calculate the costs, risks and gains of their actions, and where they find that the costs and risks of an aggressive action will outweigh the likely gains, the deterrent factor prevails. There are certain deterrence factors, like the costs of occupation are usually high, even in cases where military aggression could be carried out at a relatively low cost. The hostile reactions of other nations in the region could also be a deterrent factor. Yet, another factor of deterrence could be that, in the age of nuclear deterrence, non-nuclear factors operate as deterrents, especially outside the industrialised world.

Subrahmanyam believed that nuclear deterrence has to be differentiated from non-nuclear deterrence. He explained that in non-nuclear deterrence, the power to inflict intolerable pain on a nation would rise once its armed force is defeated. In contrast, in nuclear deterrence, the capability to inflict pain or unacceptable damage is available without a single soldier crossing a border and in a few hours.

However, he said there are certain ambiguities in the concept of nuclear deterrence; in certain cases, it is active deterrence, while in other instances, it is passive deterrence. The passive concept of deterrence is the projection of nuclear capability to dissuade the other from taking any military action and initiate a war. However, the active concept of deterrence dissuades other nations from intervening in the course of action that a nation is embarking upon to further its interests. This concept is mainly utilised in coercive diplomacy. This ambiguity makes it difficult for the analyst to explain its use for different powers.

On the viability of nuclear deterrence, KS said that it could not be judged based on either favouring or opposing the deterrence policy. In some cases, those who always supported the logic of nuclear deterrence have begun to state that holding a city and its people hostage is unethical. But this does not mean that the doctrine of nuclear deterrence should be abandoned; instead, there is a requirement to assess the doctrine and bring in specific changes within it and in its implementation. Further, in this era of nuclear parity between states, the argument that nuclear deterrence works on all occasions cannot be ensured. In a situation where both sides have enormous stockpiles of nuclear weapons, the risk of such wars getting out of hand is relatively high. The costs of such wars in terms of pain and damage are likely to be of unpredictable and unacceptable levels. This situation is applicable on both sides, and neither side is in a position to claim to deter the other without itself being deterred by the existence of nuclear weapons.

The doctrine of deterrence, according to Subrahmanyam, was based on a particular belief system. Therefore, it could not be entirely ignored as it formed the basis of defence and ideology of some powerful countries like the US and USSR. A deterrence, for Subrahmanyam, was a two-sided affair and a self-fulfilling prophecy, as those who believed that they could exercise deterrence through their nuclear arsenals were bound to be deterred by the nuclear arsenals of others. The elaborate debates of counterforce versus mutual destruction of the sixties now hold no gravity. It is just there, and it can prevent a war from breaking out, it is being argued. Yet the arms race continues to exist based on the rationale that the new weapons are needed to reinforce and strengthen deterrence and add to its capability. It has been reduced to a continuous arms race regulated only by a resource constraint and the obsolescence factors of weapons.

Deterrence is more of an operational, strategic doctrine and not an instrument of politics among nations. Therefore, the outcome is that instead of politics influencing the strategy of deterrence, the latter tends to influence international relations. Subrahmanyam explains that if it were the other way around, the face of international politics would have been different.

However, the doctrine of deterrence had got delinked from politics among nations. It had been chiefly pursued as a mechanical, operational, strategic doctrine mainly with the advancement of technology, excessive attention focused on weaponry, deployment postures and stockpiles. It took the shape of defence preparedness and a race towards arms stocking.

On India's nuclear strategy, KS cited that India had no intention of joining the club of hegemonic nuclear powers. The Indian motive is to maintain minimum nuclear deterrence as insurance against nuclear intimidation. India would become a global player and one of the centres in a polycentric world; it could then focus on its economic, technological, social and political development. He was sure India could build minimum nuclear deterrence steadily over time without indulging in an arms race. India's commitment to keeping the nuclear arsenal only for deterrence, and our pledge not to be the first one to use it, has now been reiterated by the heads of the country.

With minimum nuclear deterrence, India's policy of no first use will meet its security needs by deterring nuclear threats and guaranteeing inevitable nuclear retaliation. And therefore, to ensure the capability, India must legislate both political and military succession in command and control. KS explained that nuclear weapons are not weapons of war but weapons of mass destruction. It is a political weapon meant for intimidation, deterrence and retaliation. Further, India's willingness to join the FMCT (Fissile Material Cut-off Treaty) indicates that this country is highly realistic in assessing its deterrent needs.

Views on Disarmament, Nuclear Non-Proliferation Treaty (NPT), Comprehensive Nuclear-Test-Ban Treaty (CTBT)

The opinions on disarmament, NPT and CTBT have always been wide-ranging and have divided the world into two prominent camps. While one camp supports all three and wants the world to follow the terms and conditions laid down by a small group of powerful nations, the other camp disapproves of the policies; then there is this third camp which is in accord with the first group but not on their terms. India belongs to this third category of nations on mutually agreed conditions.

From the time of the Non-Aligned Movement, India has urged the two major blocs to restrain the development and proliferation of nuclear weapons. In the early 1950s, Prime Minster Jawaharlal Nehru called for a standstill agreement which sought to end test explosions of nuclear weapons as a first step towards discontinuing the production and stockpiling of such weapons.

In the late 1950s, the government of India expressed its reservations against the nuclear explosions. In 1961, the non-aligned nations headed by Ethiopia moved a resolution in the UN General Assembly declaring that the use and threat of nuclear weapons would directly violate the Charter of the United Nations. And any state using such weapons would be considered as acting contrary to the laws of humanity and committing a crime against humanity and civilisation. They demanded the prohibition of nuclear weapons.

The UN convened international conference recommended the establishment of denuclearised zones covering land areas and oceans of the world, notably those hitherto free from nuclear weapons, following the desire expressed by the states and people concerned.

In all summits of the Non-Aligned Movement from 1961 onwards, there was always a mention of disarmament, where India always played the role of a third-world leader. The non-aligned group has always supported the concept of a freeze on nuclear weapons as the first step in halting the continuing proliferation. India emphasised that nuclear weapons of mass destruction must be limited, reduced and eliminated wherever they existed. Hence the freeze must be followed immediately by negotiations on reducing and eliminating all nuclear weapon stockpiles.

Therefore, by pointing out the above facts and the struggles of the Non-Aligned Movement, K Subrahmanyam showed that the philosophical approach of the Non-Aligned Movement differed from that of NPT. And it was for this reason that the Non-Aligned Movement could never endorse the idea of NPT, but certainly favoured the idea of disarmament.

On the issue of export control and the north-south controversy, K Subrahmanyam, in one of his articles published in *The Washington Quarterly* in 1993, criticised the export control norms on two grounds. First, that export controls are by definition discriminatory – they embody a fundamental double standard whereby nuclear weapons and missiles are deemed essential for the security of industrialised countries but dangerous in the hands of developing nations. Secondly, export controls fail to deal with underlying problems of international security, which means that, at best, they are short-term fixes and aggravate fundamental international problems. It is bureaucratic.

According to him, export control divided the world into North and South, projected a racist bias, and has proved to be an inefficient instrument for pursuing global non-proliferation objectives.

He believed there was too much politics in the post-cold war era among the nuclear powers. There was a vast difference between what they preached and practised. And these countries made different rules for the developing world. Hence these biased behaviours of the western countries made export control a big failure.

Subrahmanyam perceived a need for new and innovative thinking to tackle the problem of nuclear weapons. One of the approaches could be to bring some changes to the existing NPT. Second, a treaty could be negotiated, and the concerns of the non-aligned nations incorporated. All nuclear installations and nuclear activities should come under non-discriminatory inspection.

Finally, on export control and the north-south issues, Subrahmanyam said that it becomes a controversy only when the counties in the north form clubs and groups and do not include the south in such arrangements.

On many occasions, KS talked about revising the NPT to include India, Israel and Pakistan or having a supplementary treaty that would incorporate all the objectives of non-proliferation and restore the term non-proliferation to its true meaning. The no-first-use provision would lay the foundation for eliminating nuclear weapons. All nuclear weapon powers would be required to disclose their nuclear stockpiles, enabling the verification agency to pursue undeclared weapons and fissile materials anywhere in the world. It would be the first step towards delegitimising and eliminating nuclear weapons. But it would not jeopardise the deterrent arsenals of those nations, that they feel they must maintain. It would eliminate the discriminatory character of the NPT. The new treaty would incorporate the above objectives and demonstrate that the world had made tremendous progress since 1968 when the NPT was signed.

On the deterrence factor in India and Pakistan, KS stated

in his article, 'An Equal Opportunity NPT' in 1993, that speaking realistically, both nations, India and Pakistan, did not require anything beyond a minimum deterrent capability. He further suggested that both the countries, after the agreement not to attack each other's nuclear facilities, can follow up by signing something like an 'Open Skies' plan and organising discussions between military staff on strategic doctrines and interactions between strategic thinkers.

Once again, through his articles, KS made a point in favour of India and said India could join the NPT but not in the 'lesser' category. There was and still exists a feeling in the Indian community that the West does not adequately appreciate the depth of public opinion in India about nuclear apartheid.

KS's writings reflect how articulate and balanced he was; he did not have a biased opinion for or against anything. He always argued and gave ample reasons to substantiate his statements from Roman history to World War history. He remembered all the instances well. He was singularly driven in his quest to advance India's interests. His forte was strategic decision-making at the national level way before it became a discipline. He had a natural ability to make a balanced decision considering the risks, rewards, and available resources. His recommendations were clear and explicit. He wrote numerous books and newspaper articles and developed a body of strategic opinion.

Family and friends reminisce that Subrahmanyam had a very sharp memory; unlike most of us who have trouble recalling a date from the past, for Subbu, it took a matter of seconds. Facts and figures were at his fingertips, and it wasn't easy to counter him. KS was encyclopaedic in his knowledge and willingly shared it. He was a realist, but his realism was based on Bhishma's teachings in the *Shanti Parva* of the *Mahabharata* and not any recent Western thinker. He was well-versed in the writings of the different authorities in international affairs and believed that they had limited applicability to the Indian situation.

Vidya Subramaniam, his colleague from *Times of India*, while sharing an interesting anecdote, said he was on first-name terms with a host of world leaders, a fact they got to know when one day he casually referred to a conversation with a certain Henry. Puzzled, they asked, 'Henry Who, sir?' and he said, Kissinger. 'It didn't occur to him that not everyone had access to the world's who's who,' recalls Vidya.

No doubt KS was a patriotic Indian, but he never mixed feelings and emotions with his vision of how a nation should be to become powerful. On one hand, he always, through his writings and interactions, disclosed the weaknesses of India and stated that the Indian government had not made adequate efforts to communicate its rationale to the world due to the absence of arms control expertise. He criticised the Indian government on several occasions. He opined that within the Indian governmental structure, there is no capability to carry out a cost-benefit analysis in matters of nuclear deterrence.

He once said that the Indian political class, foreign office bureaucracy, academia and media have yet to reorient themselves to this new situation. The traditional method of working has to change. The reactive ways have to move towards proactive strategy formulation. The initiatives on disarmament have not even obtained the support of our non-aligned friends, as seen during the NPT extension, CTBT, and the resolution to declare nuclear weapons use and threats a crime against humanity.

On the other hand, from the accounts of his daughter, we can see his patriotic and generous side, when during an Indo–Pakistan war, he contributed to the war efforts by selling off his wife's jewellery and sarees.

India-United States Nuclear Agreement

According to K Subrahmanyam, India–US relations were very much ambiguous. It was because of a certain cold-war mentality still present in significant sections of the American administration. He says,

> In their theology, India has not lined up as they expected. In any case, they do not like anybody that does not line up with them. ...we have a lot of interaction with the United States even though we don't agree with each other. That interaction is at the elite level. If you go just around (in Delhi's expensive residential areas), and ask every household if they have a son or a daughter in the United States, you will see that almost every household has one of them in the US. It is also acknowledged that, in the United States, of all the ethnic communities, the Indian community is the most affluent and the one which makes the maximum amount of contribution to US science.

> All this means that you have a very complicated relationship. People to people, there is a lot of interaction, yet at the same time we disagree on policies and issues.[9]

Subrahmanyam, in 1991, quoted Dr Henry Kissinger from his book, *Diplomacy,* about the likely emergence of a new global balance of power in the international system after the end of the cold war, comprising the US, European Union, China, Japan, Russia and possibly India.[10] Later in 1998, with Pokhran-II, India became a nuclear weapon state. India further successfully weathered away the US-imposed sanctions of 1974 and 1998. The nuclear test brought about significant changes in the world's attitude; the most positive one has been in US behaviour. The US witnessed India's capability to rise globally; therefore, its attitude towards the country soon changed, which is reflected in many episodes. The US took a U-turn on its relations with India, assessed India's potential, and called India a natural friend of the US, which, through its technological, economic and demographic interaction, will help the US sustain its pre-eminence.

According to K Subrahmanyam, a lot can be achieved from the relationship between the two nations. The US has taken the initiative, and it is now India's turn to respond. Still further, he explains that the culture of assessment is in long and

9 Berthier Serge, January 2000 'Interview with K. Subrahmanyam', *Asian Affairs Magazine,* http://www.asian-affairs.com/India/subrahmanyam.html.

10 Subrahmanyam.k, September, 2005, 'Indo–US Relations in the Coming Decades', *Journal of United Service Institution of India.*

short-term relations. He mentions how long back, Nehru said that national interests and not ideology determine a nation's foreign policy and international relations. Therefore, India should adopt Bhishma's wisdom when he told the Pandavas, the idea of statecraft should be based on the thought that 'for the king, there is no friend or enemy as such, but circumstances made friends and enemies. Hence India should be assessing what circumstances have brought about this change of attitude in the US policy instead of rejecting the US initiative outright or accepting it uncritically.'[11]

In line with his *Realpolitik*, Subrahmanyam, once a well-known critic of the US for its earlier Cold-War-influenced foreign policy, even being dubbed a 'leftist' and 'Soviet sympathiser' on occasion, later on became a leading advocate of the Indo–US Nuclear accord on civilian nuclear cooperation signed by President George W Bush and Prime Minister Manmohan Singh in 2008.[12] Subrahmanyam was critical in saving the India–US civil nuclear energy cooperation agreement in March 2008. After months of trying to get the Left Front on board, the Congress Party had virtually given up on the deal.

Subrahmanyam saw the India–US civil nuclear energy agreement as a step towards eliminating the high-tech barriers the US had erected against India so that the two democracies could work together in the knowledge economy. As a strategic

11 Ibid: 370.

12 Bhatt Sheela, 2005, 'No American can treat India like a pet', the interview, http://in.rediff.com/news/2005/oct/11inter.htm.

visionary, Subrahmanyam recognised early that economic competitiveness and technological competence lie at the heart of national power in the modern world (Sanjaya Baru, *Business Standard,* 2013).

He further opined in his article in *India Today* that 'considering the positive reactions of several industrialised nations towards making an exception for India with regards to the Non-Proliferation Treaty (NPT) for civil nuclear cooperation, it is obvious that the move is not a conspiracy by Washington to trap India and disarm its meagre nuclear arsenal but a major international effort to enable India to grow fast as an economic and technological power so that it can contribute to an Asian and global balance of power.'[13]

Later through many interviews and articles, he attempted to explain his changed stances vis-à-vis America, and that India should use this excellent opportunity to work with the US. He argued that nothing is permanent in international relations, which validates what Bhishma told Pandavas from his deathbed – that circumstances determine friends and enemies. Therefore, what is good for India is based on changing realities, without loyalty to any ideological fundamentalisms, religious or secular.

He, thus, concluded that although the Americans had to abide by their national procedures, the deal was much beyond a bilateral one and involved the global community. Hence, it

13 Subrahmanyam. K., December 2006, 'Ending the Isolation', Indiatoday.in, http://indiatoday.intoday.in/story/k.-subrahmanyam-is-a-former-secretary-to-the-government-of-india-and-a-strategic-affairs-expert-on-indo-us-nuclear-deal/1/180040.html

would have been an unpardonable folly to reject this opportunity.

Thus, through his articles, Subrahmanyam at the time played a vital role in reviving the issue and encouraging the Government to take a tough stand.

Conclusion

K Subrahmanyam was indeed the initiator of strategic discourse in India. He played a pioneering role in making this country conscious of the need to think about national security and do so strategically in a milieu where the very concept of strategic studies was absent. Some called him 'Bhishma Pitamaha' of the expanding circle of young scholars on strategic thinking.

Stephen P Cohen once said, 'Subbu, the most important strategic thinker of India, is also one of the handful of the world's truly original writers and speakers in this arena.' He further said that 'no other Indian had the same impact on American officials, scholars and strategists' as KS had. Even those in the US who considered KS anti-American and worse, respected his 'formidable intellect, unequalled mastery of facts and sense of humour'.

KS's article in 1998 on 'Nuclear India in Global Politics' had created a sensation; he did not merely analyse the security challenges India faced but boldly set out his policy prescriptions. He was marked by his daunting knowledge of facts and relentlessly logical analysis, which was his hallmark for several years. His ability to take an Indo-centric view of the changing world scene and its inevitable impact on this country without ever being chauvinistic adds lustre to his work.

Many people had criticised KS for his stance towards the US – from once being principally against the United States in the past to becoming soft on America. But the fact is that he continued to remain against US unilateral policies. However, he tried finding positives in the US behavioural change, and believed India should understand and use these opportunities to its advantage.

'He declined the trappings of power to work in a think tank,' said close associate and head of the National Maritime Foundation, C Uday Bhaskar, referring to Subrahmanyam's two terms as head of the Institute for Defence Studies and Analyses (IDSA) – in 1968-75 and 1980-87. 'He was the father of India's nuclear thinking.'

BG Verghese, another close associate and former editor of the *Hindustan Times* and *The Indian Express*, said that during his association with IDSA, Subrahmanyam 'nurtured generations of scholars'.

Narendra Sisodia, IDSA's former Director General, described Subrahmanyam as an 'intellectual genius'. 'From being considered pro-Soviet, he became one of the biggest proponents of the Indo–US partnership. He was also influential when it came to the (landmark 2008) "Indo–US nuclear deal" that has seen the ties between the one time "estranged" democracies warm to the level of "strategic partners",' Sisodia added.

Tarun Das, former Chief Mentor CII (Confederation of Indian Industry), who was part of the PM's Task Force on 'US global strategy: Emerging trends and long-term implications', chaired by Subrahmanyam in 2005/2006, reminiscing about his relationship with Subrahmanyam writes, 'I was part of his seven-member team. His leadership was just amazing. He went into history to help each of us understand US strategy over decades and, then, to discuss and determine what India's strategy should be on each key issue – defence, economy and trade, energy and technology.'

'Another area he influenced enormously was the whole process of Track-2 Dialogue, even though he, personally, could only participate once in Jaipur. He taught us the value of Track-2, the ability to speak freely, to listen to equally frank views on India from our counterparts, and the impact of all of this on building mutual, shared, understanding.' Tarun Das adds.

According to Das, Subrahmanyam believed that India must not be isolated. India must work in partnership. And an essential partnership, in his view, was with the US because of India's own national interests in defence, economy, energy and technology, which always led India to the USA.

Apart from being a serious writer, KS sometimes wrote humorous columns under different names like *'Ambi'*, *'Atreya'* and *'Mani'*.

One of his most famous articles titled 'Agni Pariksha' was published in *Times of India* under the pseudonym KS Ambi. In this article, narrating his visit to a restaurant in South Delhi with his grandchildren, he writes how after Dr Kalam was made the President of India, this restaurant, where Dr Kalam was a regular visitor, decided to honour him by revising their menu. For instance, their Presidential Executive Thattu (thali) had three varieties – Agni I, Agni II and Agni III, wherein

Agni III was the hottest, Agni II was medium hot, and Agni I was just hot.

On a personal front, KS was a man of great values; he always respected women, which he learnt from observing his mother and her contribution to his upbringing and running of the household. Sulochana Subrahmanyam, fondly remembering her husband, said that although he never had an interest in music, he always encouraged her to pursue her PhD in music. Once, she recalls, she was down with a fever. She had to submit her MPhil dissertation. KS went and got her dissertation printed and bound; he later went to submit the dissertation as well. She further praised him and said he never imposed any decision on her and let her choose whatever she loved doing and wanted to do. He always told her that they were blessed and should be proud to have such children.

Subrahmanyam was fond of Lata Mangeshkar and MS Subbulakshmi's songs. Apart from Indian music, he also loved listening to Laura's Theme from Doctor Zhivago and singing along with it.

KS had some unique habits for which he was very famous amongst his family and friends. One was his love for various kinds of pickle; his dining table was always full of bottles and jars of different varieties of pickles and chutney.

Another unique trait was his punctuality. He considered punctuality the greatest virtue and reached everywhere on time. Unfortunately, he often arrived too early at functions and parties. He would be there even before the *shamiana* (flat ornate tent) had been put up and the chairs set out. His daughter recalls an incident when he drove off because her mother was a few minutes late getting to the car.

Another popular punctuality story was of an incident that occurred in his early days as a government officer. He had his own key to the office since most days he reached office even before the *chowkidar* (watchman), cleaners and the rest of the staff arrived. It was a rainy day. He was quite wet when he reached the office, so he took off his shoes and shirt in an attempt to dry them. He sat down on the floor to start working on a stack of files. He had a meeting in the morning with someone reasonably important. He was immersed in his files and lost track of time. At some point, someone knocked on the door, entered, and saw him on the floor. The person asked if he was the only one there, and KS said yes. The person retreated. At some point, Subrahmanyam realised that it was late and got up to put on his shoes and shirt. He got ready for the appointment, which was with the same person who had entered earlier. One can only imagine this person's shock, now seeing him in the boss's seat.

At one stage, he took a fancy to Delhi Emporium kurtas; then there was the time he turned up all spiffy in a three-piece pin-striped suit to meet his daughter's colleagues from the US, who were dressed in casual jeans. All said, at home, he always remained simple and ever busy with his work and favourite channel, BBC.

These accounts of his family and friends give us a glimpse into the life of this great person and his amazing simplicity, which made him everyone's favourite. On one hand, he impressed the world with his knowledge and powerful arguments; at home, he won the hearts of his family with his innocence and quiet ways of showing his love and affection for them.

REFERENCES

- Berthier, Serge, 'Interview with K. Subrahmanyam' *Asian Affairs,* http://www.asian-affairs.com/India/subrahmanyam.html, 2000.
- Bhatt, Sheela, 'No American can treat India like a pet', *National Affairs,* URL:http://in.rediff.com/news/2005/oct/11inter.htm, 2005.
- Event Summary, 'The career and Ideas of K. Subrahmanyam', *Brookings,* URL:http://www.brookings.edu/events/2011/02/18-india-subrahmanyam.
- Ghosh, Amitav, 'A Million Pound Note…' Interview with Amitav Ghosh in 1998, *Outlook,* URL: https://www.outlookindia.com/website/story/a-million-pound-note/215502, 2022.
- Hindustan Times, 'K. Subrahmanyam (1929-2011) – the Dean of Indian Strategy', *Hindustan Times,* URL: https://www.hindustantimes.com/delhi/k-subrahmanyam-1929-2011-the-dean-of-indian-strategy/story-GkmGw5WB5HWr3wGBHAXfiL.html, February 2011.
- India Today, '50 Power People 2006', *India Today,* URL:https://www.indiatoday.in/magazine/cover-

story/story/20060320-50-most-power-people-in-india-783469-2006-03-19, 20 March 2006.

- India Today, 'India's High and Mighty: From Corporate Czars to Policy Mavens', *India Today*, URL: https://www.indiatoday.in/magazine/cover-story/story/20060320-50-most-power-people-in-india-783469-2006-03-19, 2006.
- Jaishankar Dhruva, 'K. Subrahmanyam (1929-2011), *Outlook*, URL: https://www.outlookindia.com/website/story/k-subrahmanyam-1929-2011/270326, February 2022.
- Jaishankar, Dhruva, 'Tribute: K. Subrahmanyam', (1929-2011), February 2011, *Outlook Magazine*, URL: http://www.outlookindia.com/article/k-subrahmanyam-19292011/270326, 2011.
- Metcalf, Roy, 'A Farewell to India's Henry Kissinger', *Foreign Policy*, URL: https://foreignpolicy.com/2011/02/03/a-farewell-to-indias-henry-kissinger/, February 2011.
- Perkovich, George, *India's Nuclear Bomb: The Impact on Global Proliferation, University of California Press*, 2002.
- Ramana, 'K. Subrahmanyam: A Tribute', *Bharat Rakshak*, URL:http://forums.bharat-rakshak.com/viewtopic.php?p=1031338, February 2011.
- Ramdas, L., 'Arrogant Nuclearism', *Frontline Magazine*, URL:http://www.frontline.in/static/html/fl1621/16211020.htm, October 1999.
- Singh, Shubham, 'India's Nuclear Doctrine and Debate of No First Use Policy', *SSRN*, URL: https://dx.doi.org/10.2139/ssrn.3450297, September 2019.
- Subrahmaniam, Vidya, *Farewell to Bomb Mama*, URL:http://

vidyasu.blogspot.in/, February 2011.

- Subrahmanyam, K., 'Nuclear Deterrence', in *India and the Nuclear Challenge* (ed.), Lancer International Publication, 1986.
- Subrahmanyam, K., 'Role of Non-Aligned States in Efforts to Reduce the Nuclear Danger', presented at 36th Pugwash Conference on Science and World Affairs, Budapest, 1986.
- Subrahmanyam, K., 'Export Controls and the North-South Controversy', *The Washington Quarterly:* Vol. 16, No. 2, p. 135-144, 1993.
- Subrahmanyam, K., 'Ending the Isolation', *India Today,* URL:http://indiatoday.intoday.in/story/k.-subrahmanyam-is-a-former-secretary-to-the-government-of-india-and-a-strategic-affairs-expert-on-indo-us-nuclear-deal/1/180040.html, December 2006.
- Subrahmanyam, K., 'Warm up to the Future', *The Indian Express,* URL:http://archive.indianexpress.com/news/warm-up-to-the-future--------/8329/0, July 2006.
- Subrahmanyam, K., 'Nuclear India in Global Politics', *World Affairs: The Journal of International Issues,* Vol.2(3), July-September 1998.
- Subrahmanyam, K., 'Elimination or Irrelevance: Arms Control Today', *Arms Control,* URL:http://www.armscontrol.org/act/2008_06/Subrahmanyam, June 2008.
- The Financial Times, 'Vajpayee Opposed Nuclear Option in '79', *The Financial Times,* URL:https://www.financialexpress.com/archive/vajpayee-opposed-nuclear-option-in-79/97729/.
- TNN, 'Those who said no to top awards', *The Times of*

India, URL:https://timesofindia.indiatimes.com/home/sunday-times/deep-focus/those-who-said-no-to-top-awards/articleshow/2714586.cms, 20 January 2008.

- Whiteneck, Daniel, 'The Industrial Revolution and Birth of the Anti-Mercantilist Idea: Epistemic Communities and Global Leadership', *Journal of World-Systems Research,* Vol 2, No. 1, URL: https://web.archive.org/web/20080705140426/http://jwsr.ucr.edu/archive/vol2/v2_n1.php, 1996.
- Xavier, Constantino and Cohen, Stephen, 'The Career and Ideas of K. Subrahmanyam', *Brookings,* URL: http://www.brookings.edu/events/2011/02/18-india-subrahmanyam, 2011.

Krishnaswamy Subrahmanyam

JASJIT SINGH
The Nuclear Strategist with The Red Turban

Manpreet Sethi

AIR COMMODORE JASJIT Singh AVSM, VrC, VM (retired) was amongst India's most versatile strategic thinkers whose understanding and writings traversed the expanse of subjects like aerospace and air power, challenges posed by Pakistan and China, higher defence organization, India's foreign policy, defence economics and modern forms of warfare, as well as social issues impacting national security, counter-insurgency and many more. He received the Padma Bhushan in 2006 for his outstanding service to the nation in the field of defence and strategic affairs. However, particularly close to his heart was working towards India's nuclear future.

The role that nuclear energy could play in India's energy mix, and the role that nuclear weapons should play in national security strategy were both issues on which Jasjit Singh spent a lot of time – thinking as well as writing. On the former issue, he framed the need for nuclear energy in the context of India's energy security and argued in favour of the country's accommodation into the international nuclear regime so that outside help could be available to the indigenous nuclear power programme for its rapid expansion. On the latter issue of nuclear weapons, he was clear that it was a necessary evil given the security compulsions of the neighbourhood. But he never gave up the idea that India must continue to press for universal elimination of nuclear weapons. In fact, he had plans to write his next book called *Counterstrike,* to further the understanding of No First Use (NFU), a much misunderstood concept in India's nuclear strategy. Global acceptance of NFU, he firmly believed, could also serve as a stepping stone towards universal nuclear disarmament. Unfortunately, his book was not to see the light of day. He passed away on 4 August 2013 after having lived and breathed national security for over six decades of his active service – almost half of which was spent in think tanks.

Jasjit Singh was a man who donned many feathers in one cap. Not only did his cap, in this case, the all-time red turban that he wore, become a part of his personality that still flashes vividly in fond memory, but it also reflected the simplicity of his life. During the times when his mind was not occupied with bigger and vital issues like nuclear strategy and foreign policy, he claimed in jest how his red turban matched most of his wardrobe. Over time, it became a part of his impeccable dressing style; it etched an image of Jasjit Singh that stayed on with everyone who knew him for the man he was.

Early years

Born on 8 July 1934, in his ancestral home in Chandpur Rurki, a small village close to Nawanshahr in Punjab's Shaheed Bhagat Singh Nagar district, Jasjit Singh was one of eight siblings. Born into a landed family, his father felt that Jasjit's first responsibility lay in helping run the farm, even though he graduated with English (Hons) from a college in Shimla. Jasjit Singh was always very proud of his proficiency in the English language, which he acquired first from the missionary school that he attended in Lahore and then honed during graduation. One is often reminded of his insistence on the distinction between certain words such as *will* and *shall* – vocabulary that is pretty much used interchangeably today. But he would always remember his school grammar that had ingrained in him the difference in usage of the terms. He brought this to bear in the drafting of the nuclear doctrine where he was sure of why he was writing 'India shall pursue a doctrine of credible minimum nuclear deterrence' or that 'any nuclear attack on India and its forces shall result in punitive retaliation with nuclear weapons to inflict damage unacceptable to the aggressor.' For him, *shall* spelt a definitiveness that brooked no hesitation, whereas *will* could indicate the exercise of an option.[14]

After finishing his graduation, Jasjit wanted to join the Indian Air Force (IAF), but surprisingly enough, his father,

---

14 Modern English usage of the two terms can be found in the sentence, '*Will* can imply volition or intention, while *shall* can imply necessity', as cited in Maeve Maddox, 'The Difference Between "Will" and "Shall"', https://www.dailywritingtips.com/the-difference-between-will-and-shall/.

who had himself retired as Chief Minister of Nabha State and returned to farming, disapproved of the decision. As stated earlier, he wanted Jasjit to join him in farming, and the son complied. But a period of drought during 1951-53 convinced him that he could not make a career out of farming. Finally, his father agreed to let him join the IAF and Jasjit joined the service in 1954, earning his commission in 1956.

Soon after, Jasjit Singh graduated with top honours and was awarded the much-coveted Jodhpur Sword of Honour. He served the institution of IAF for thirty-one years with great distinction. As an ace fighter pilot, not only was he an extremely qualified flying instructor but also tenanted several operational, command and instructional appointments. His participation in the Indo–Pak war of 1971, where he carried out air attacks that destroyed Pakistani tanks, won him a Vir Chakra for his gallantry in the face of enemy action. Subsequently, in 1974, he was awarded the Vayu Sena Medal by the IAF for his exceptional devotion to duty, exemplary leadership and conduct. The Ati Vishisht Seva Medal came his way in 1984 for his distinguished service. Indeed, Jasjit Singh's cap had begun to be adorned with many feathers by the 1970s itself.

During his time as a fighter pilot, Jasjit underwent two ejections from aircrafts he was flying. The first such incident took place on 18 October 1973 (which was also his wedding anniversary) when a bird hit caused an engine failure and led to loss of control over the aircraft. The mishap happened while he was posted at Hindon. The second incident took place in 1977 when he was commanding 17 Squadron at Halwara. This time the aircraft had an engine malfunction and he was compelled to eject. Both times, during the ejections, he suffered multiple spinal injuries, but his destiny brought him back for bigger things in life. Many a time, and especially when he held on to contrarian positions, he would joke that since he had broken his spine twice, it went to show that he had a spine!

By the early 1980s, once it was clear that the injuries would not let him be fit enough to fly fighter jets again, his academic interests came to the fore. He completed his Masters in history from the University of Mysore during this time. He also started writing a series of articles on air power for *Strategic Analysis*, the flagship journal of the then only national security think tank in India, the Institute for Defence Studies and Analyses (that was rechristened Manohar Parrikar IDSA in February 2020) at New Delhi. This is how he and IDSA got introduced to each other and their destinies were to remain intimately intertwined for sixteen years thereafter.

The journey as a strategic thinker

In 1985, Jasjit Singh became the first serving military officer to be appointed Deputy Director at IDSA. This was an unprecedented move both for the IDSA and for the military. He himself used to say that few in the IAF understood his decision at the time. They considered it crazy for him to be giving up a promising service career to 'move to the side-lines' to do research on air power and national security. At IDSA, the Director then was K Subrahmanyam, the individual who India remembers as the doyen of national strategic thinking, particularly on nuclear strategy. The two came to form a formidable team as they wrote and spoke on matters of national security.

Two years later, in 1987, K Sub, as he was called by all, decided to retire from his active position as Director. He reposed his faith in Air Cmde Jasjit Singh to step into his shoes. Many years later, Jasjit described this moment as having arrived too quickly for him. He found the shoes of Subrahmanyam too large for him; he felt that he had only been shuffling along in them! However, for fourteen years after the change of guard at IDSA, Jasjit remained in the position of Director until 2001. He made a huge contribution towards earning such a stature for the organisation that made it count across the world as one of the leading policy think tanks. The crop of scholars he groomed and mentored as Director, today constitute the vibrant strategic community that populates nearly every think tank in India.

The dawn of the 1990s came with the end of the Cold War. As the US–USSR/Russia relations transformed, it spelt a tectonic shift not just for the world in general but also for India. The collapse of the USSR, with whom India had forged a deep strategic relationship, and the brief unipolar moment with the US as the sole superpower, with whom India had not enjoyed great relations during the Cold War, put New Delhi in a difficult diplomatic situation. At this juncture, Jasjit Singh stepped up his efforts to secure the country's interests in a changing geopolitical landscape. While he engaged with his counterparts across the world, of particular relevance were his interactions with USA,

Russia and China. In fact, during the decade of the 1990s, he led eight rounds of Track II Indo–US strategic dialogues. He also annually addressed military academies in Moscow and Beijing. These engagements had a major impact on shaping perceptions about India, and in imparting an understanding of India's security challenges. Further, this served as a period of laying down the groundwork which had a huge role to play in the quick turnaround in American perceptions and position on India after the conduct of nuclear tests in 1998. He and Subrahmanyam (who remained a frequent participant at IDSA events) made a significant contribution to shaping the country's nuclear discourse at a time when it was ostracized and under severe international sanctions.

An anecdote comes to mind about the man who inspired many scholars and was instrumental in harnessing the potential of many young minds. In 1997, I had already submitted my doctoral thesis at the Jawaharlal Nehru University, New Delhi, and was eager to start a career as an international affairs research analyst. Air Cmde Jasjit Singh, then Director of IDSA, interviewed me. He did not check my academic credentials and declined to see my certificates. Rather, I remember being asked by him to spell out my dreams for myself and my country. After a few minutes of conversation on the state of nuclear non-proliferation, he asked me, 'So when do you want to join? Tomorrow?' This was typical of the man – ever welcoming of young scholars and always willing to give them a chance. He never dithered in appointing new scholars. He used to joke that people called him a 'tin pot dictator' for making these decisions without calling for an appointments committee and following elaborate procedures. But, he took these decisions from his gut instincts, and it is clear that they worked well, since there are many in the strategic community today who can attribute their careers to the platform he gave them. I was lucky to be one of them.

It was also my good fortune to join the IDSA in 1997, a year before India tested its nuclear weapons. The momentousness of the event and its implications revealed themselves to me through my observations of Jasjit Singh's myriad interlocutions on the subject with others of the Indian and international strategic community, as well as through my many interactions with him over the years. As it may be recalled, when India tested its nuclear weapons on 11 and 13 May 1998, the international community led by the USA and its allies erupted in anger and strongly criticized the action. Sanctions and technology denials followed as the American President Bill Clinton led the demand that India 'cap, roll back and eliminate' its nuclear weapons programme.

At a delicate time like this, India's diplomatic cadres went into overdrive across all major capitals of the world to explain the nation's nuclear compulsions. As the Director of IDSA, India's premier national security think tank that was beginning to be recognized and respected around the world, it also fell upon Jasjit Singh to put his weight behind this effort. He contributed wholesomely to the making of informed opinion on India's nuclear policy. He travelled across the length and breadth of the world to explain Indian threat perceptions to policy makers and opinion shapers and thus subtly influenced their view of India. It is hardly surprising that so many in the strategic world remember him as the man with the red turban who passionately, but politely, argued the case for India's nuclear status.

The period 1998-2001, the four years before he retired from IDSA, could perhaps be counted amongst the busiest ones of his career as a strategic analyst. He spent nearly all his time addressing conferences across the world, participating in TV discussions, writing short and long articles, hosting delegations from many countries, organizing conferences. Every action was a brick in creating an understanding of what a nuclear India meant.

Meanwhile, just before the nuclear tests in 1998, the National Democratic Alliance (NDA) Government, in fulfilment of one of the agenda points from its election manifesto, set up a three-member Task Force under KC Pant to review policy challenges and recommend structural reforms to India's higher defence organisation. Jaswant Singh was a member of the Task Force, while Jasjit Singh was made its Convener. In this capacity, he laboured for months and temporarily also relocated himself to an office on Parliament Street as he immersed himself in the exercise. The group managed to produce a comprehensive document that recommended measures that could be taken to remedy the situation. These steps eventually led to the creation of the National Security Council (NSC), Strategic Policy Group (SPG) and the National Security Advisory Board (NSAB), even though the NSC model that was adopted was not truly in concurrence with the one recommended by the KC Pant Committee.

Even as Jasjit Singh was busy with the task mandated to the Committee, I distinctly remember the afternoon when he summoned some of us, the IDSA researchers, to his new den to share his idea of bringing out a book, the name of which he had already thought of – *Nuclear India*. He wanted to document the new and significant development in India's nuclear journey. Each one of us was tasked to write a chapter for the book and we were given a stiff deadline. Consequently, in only two months of India going nuclear, the book was published to explain this new reality to the world and to the people at home. The volume explained the need for India to conduct the nuclear tests at that juncture, the complexities that the new status would bring for the country, and the doctrine and strategy it must follow as a path towards an affordable credible deterrence. To this day, the book remains an essential reading for anyone wanting to understand nuclear India.

A peep into Jasjit Singh's nuclear thinking

Jasjit Singh was clear that the reason for India's acquisition of nuclear weapons was not prestige or status, but security. The former, he firmly believed, would only come from 'how we solve our problems and how we conduct ourselves in the face of evolving geo-strategic and geo-economic realities' (Singh, 1998). Nuclear weapons, on the other hand, were meant to provide an 'assured environment of peace and stability in which our primary goal of human development can be pursued unhindered.' Therefore, in his reckoning, the weapons were an 'insurance against challenges to peace, stability and our independence of decision making' (Singh, 1998: 10). And, the environment so created was to be exploited by the country to grow and develop. Prestige, he believed, would naturally follow.

In his assessment, the biggest threat to India's rise was the uncertainty posed by China's ascendance to economic and military power and the future direction of its policies. The India-China bilateral agreement of 1993 (on Maintenance of Peace and Tranquility along the Line of Actual Control) was based on the principle of 'mutual and equal security', and in Jasjit's reckoning, the acquisition of nuclear weapons strengthened India's 'equal security', which in a case of nuclear asymmetry would have been a mirage. He wrote, rather presciently, in 1998:

> China is, and is likely to remain, the primary competitor and challenge to India in strategic terms, ranging from competition for investments, markets, political influence, especially in the developing world and in possible military terms.
>
> (Ibid: 19)

This sentence written two decades ago is today a reality as China flexes its military muscles in South China Sea and reaches out to developing nations across the world with its Belt and Road Initiative (BRI).

While explaining that the centrality of China as a security threat was the reason for India to acquire nuclear weapons, Jasjit Singh, nevertheless, also took pains to explain that before taking the difficult decision of conducting nuclear tests, India had long pursued a policy of keeping its nuclear options open. 'This was an approach in keeping with Indian civilizational value of taking the veritable middle path,' said Jasjit Singh (1999:16). India kept to this path in the hope that the world would move towards disarmament at some stage and that would address India's security concerns from China. But the move of the world at the end of the Cold War from disarmament towards a renewed focus on discriminatory non-proliferation (with the extension of the Non-Proliferation Treaty in 1995 and the conclusion of the Comprehensive Test Ban Treaty in 1996 as well as strengthening of ad hoc export control instruments) put pressure on India's ability to stay on the middle path. As he pithily put it:

> The situation that India faced by early 1998 was that if the stranglehold of the non-proliferation order continued to tighten around the open option, and India did not take steps to break out of it, very soon there would be no option left.
>
> (Ibid: 25)

So, it was hardly surprising that India decided in favour of testing, willing to bear the costs the decision might entail, over living in a state of permanent nuclear adverse asymmetry.

Despite having taken the step of possessing nuclear weapons, Jasjit Singh was clear about India's ethical stand. He wrote:

> India's strategic and security interests are served better if there are no nuclear weapons that can impinge on India's security calculus. Nuclear disarmament, therefore, is not only a moral/

> ethical principle for us, and a necessity for international peace and security, but also an imperative for national security.
>
> (Ibid: 289-90)

According to him, this factor had not altered even after India went nuclear, and, in fact, could even be leveraged to 'strengthen our negotiating position to seek disarmament'.

As a vocal member of the international Pugwash Movement that argues for peace and security through the correct use of science, Jasjit argued passionately in favour of universal, multilaterally negotiated, verifiable disarmament. In 1988, he was pivotal in the drafting of the Action Plan that was presented by then Prime Minister Rajiv Gandhi at the Third Special Session on Disarmament. Twenty years later, in June 2008, he planned and executed (despite having fallen seriously ill in February that year) an international conference – 'Towards a Nuclear-Weapons-Free World' – which was addressed by the Prime Minister and the Vice President of India and was attended by as many as two hundred members of India's strategic community, including fifteen international experts. He had conceived the idea of the conference towards the end of 2007 and had it vetted by then National Security Advisor (NSA), who agreed readily, including providing funding from the Ministry of External Affairs for the event. The Indian Council of World Affairs (ICWA) also agreed to co-sponsor the conference and share the myriad tasks as we went along.

Jasjit Singh wanted it to be an international conference. He quickly set into motion the process of contacting his vast network of strategic friends spread across the world. However, before the implementation of his ambitious plans could commence, he suffered a severe brain haemorrhage in February 2008 and remained in hospital for several weeks. Through March and April, the state of his health caused much concern, even as the Additional Director, AVM Kapil Kak valiantly held fort. Soon after his discharge though, while not allowed to come to office, he made sure that the office came to him! So, from his study, he directed the execution of the humongous task of crafting the programme, sending out invitations, overseeing flight and hotel arrangements and ensuring the smooth conduct of the conference. The prospects and challenges of universal nuclear disarmament were well examined in the two-day deliberations inaugurated by the then Prime Minister. As explained by Jasjit himself in his introductory remarks:

> We plan to explore the nuclear environment of the future in its diverse dimensions and discuss the ways and means of dealing with them… As we know, consciousness of the danger is a prerequisite to our ability to deal with it successfully. It is the consciousness slipping back in our minds that led to lowered attention to abolition of nuclear weapons after the end of the Cold War.
>
> (Sethi, 2009:12)

He also made a persuasive case for 'a paradigm shift from competitive security to cooperative security' as a way of managing future challenges specially to manage international peace and security in a non-nuclear world. The papers presented by scholars of high stature made for a worthy volume and it was the big-heartedness of the man that he asked me to be the editor of the book that was published in 2009 (Sethi, 2009).

Despite being a supporter of universal nuclear disarmament, Jasjit had no illusions that such a world was going to come about either quickly or easily. So, while he exhorted India and other countries to make efforts to achieve this goal someday, for the present he was realist enough to understand that India would need to possess nuclear weapons to protect its national interests. He recognized that this also meant drafting a clear and unambiguous doctrine having an appropriate role for nuclear weapons. This task too was ably performed by the Air Commodore in his capacity as a member of the first National Security Advisory Board constituted by the Government of India under the chairmanship of K Subrahmanyam. Along with other astute minds well versed in understanding of nuclear deterrence, as part of the group, the duo worked hard to put together a draft doctrine. This was produced in record time and presented to the government around the middle of 1999. The draft, in a rather uncharacteristic move for India, was made public by then National Security Advisor Brajesh Mishra in August 1999.

Contours of what came to constitute India's nuclear doctrine can be discerned from Jasjit's chapter on 'A Nuclear Strategy for India', in *Nuclear India* that was published almost one year before the doctrine was drafted. He opined in the chapter that India's nuclear doctrine must take its direction from the logic of the role that India wants to bestow upon its nuclear weapons. In his view, the role of India's nuclear weapons was to dissuade and deter nuclear blackmail and coercion and to prevent the threat of use or use of nuclear weapons against Indian territory or people (Singh, 1998). With this in view, he found no role for nuclear weapons for war-fighting. Accordingly, he firmly ruled out the need for any pre-emptive use of the weapons. With NFU as the driving strategy, he rightly put emphasis on survivability of the nuclear arsenal as a crucial necessity. As he pointed out, 'various ways will have to be found so that an adversary is not tempted to believe that a first strike against us will seriously degrade our ability to retaliate' (Ibid: 314).

Writing on the issue of tactical nuclear weapons, a capability that Pakistan introduced into its nuclear deterrence posture with its first test of the very short range ballistic missile, Nasr, in April 2011, Jasjit had already dismissed the concept of tactical nature of nuclear weapons in 1998, especially between physical neighbours. He wrote:

> Any nuclear weapon of any quality, mode of delivery or yield, used against any type of target, will result in a strategic impact to which the logical response would be the use of nuclear weapons, more often than not, on an overwhelming scale… Regardless of the yield or type of delivery system, we must ensure that we do not become the victim of believing in the fallacy of any nuclear weapon being tactical. The artificial division of nuclear weapons into tactical and strategic is not only irrelevant for us, but carries with it the danger that a belief system could grow in a way that might justify the use and utility of such weapons for actual war-fighting.
>
> (Ibid: 317)

As is evident, the book *Nuclear India* sought to make public the altered context and content of India's nuclear debate. His attempt was to highlight the issues that would need serious and detailed consideration and evaluation in the public domain so that a national perspective could emerge. He himself acknowledged that the book:

> [S]et forth some ideas, not with the belief that they represent a finality of logic and necessity, or even to perceive them as some sort of prescription, but with the hope that they will trigger further discussion and debate to evolve our nuclear doctrine and strategy.
>
> (Ibid: 323)

Indeed, some of the principles and beliefs that he then articulated have stood India in good stead as its capabilities have improved in keeping with the guidelines provided by the doctrine. In the closing pages of the book, Jasjit Singh made an assertion. He wrote:

> War in the sub-continent, if it takes place, will remain a 'border war' limited in time, space and aims. The 'N'-factor will continue to weigh heavily on the prospect, scope, and nature of a conventional war involving India... in the past, the wars of the subcontinent were limited (in time, scope, goals, etc.) by choice. But nuclearisation has made wars limited as an imperative.
>
> (Ibid: 311)

This proved to be so uncannily correct when within a

year of the publication of this book, Pakistani regular soldiers infiltrated into Kargil dressed as *mujahideen*. This was a bold step taken by Pakistan under the assumption that its nuclear weapons gave it the freedom to undertake offensive actions with no risks since the international community was expected to step in quickly to disentangle the situation. As Jasjit often explained, the central logic of Pakistan's acquisition of nuclear weapons was to 'neutralise Indian conventional superiority to the degree that any punishing military operation by India in retaliation to its actions is forestalled by the risk of escalation to a nuclear weapon exchange' (Singh, 1999:125). However, while conceding that the presence of nuclear weapons may have made full-scale wars difficult, Singh was a votary of the concept of limited war. In fact, he believed that the country that could manage the dynamics of limited war would be able to run a successful national strategy.

In Kargil, as India fought back to regain its territory, it was able to 'demonstrate the politico-military acumen to conduct a limited war' (Ibid: 126). This negated the Pakistani assumption of an early involvement of the international community that would insist on a ceasefire, thus enabling salami slicing of territory. But, signs of definite restraint by India to neither expand the scope of conflict by opening another front, nor allowing its aircraft, engaged in the operation of dislodging the Pakistani soldiers, to cross the Line of Control (LoC), led the international community to tilt its opinion in favour of India. The ability to execute a limited conventional response when both sides had demonstrated nuclear capability was seen as responsible behaviour, and it stood out in stark contrast to the irresponsible and provocative actions taken by Pakistan.

Pakistan's defeat at Kargil created considerable misgivings between the civilian administration and the military institutions. After a number of developments that heightened the tensions, Prime Minister Nawaz Sharif decided to remove then Chief of Army Staff (COAS), General Musharraf, from the position. This act was done when the COAS was not even in the country and is reported to have received the news as he was playing golf in Sri Lanka on the fiftieth anniversary of the Sri Lankan Army. The Pakistani Army, however, could not have been expected to take this move lightly. In fact, it was a matter of little surprise that on 12 October 1999, with the support of the Chief of General Staff, General Musharraf managed to stage a coup and ousted the Prime Minister. He even elevated himself to the position of Chief Executive of Pakistan.

The day this event took place happened to be one on which IDSA was hosting a delegation from abroad. At the evening dinner organized for them, on the India International Centre (IIC) rooftop, the talk was all about the latest development in Pakistan and its future as a nuclear state with the army in the saddle. Little did I, then a fledgling in these circles, realise the importance of what was being discussed by the stalwarts at the gathering. But, this was a unique way of mentoring by the Air Commodore. He gave ample opportunity to young scholars to rub shoulders with seniors and encouraged them to ask questions, sit at the main table and engage in conversation. It was a sheer delight to see him in action at such events –

a gracious host and a firm speaker upholding India's national security interests.

An institution builder - from IDSA to CAPS

After retiring from IDSA at the age of sixty-seven, Jasjit Singh briefly took on the role of editorial advisor for defence and strategic affairs at *The Indian Express*. Enjoying a great rapport with the young work force at the newspaper office, he was equally at ease in the small cubicle. He had a kettle and a couple of cups tucked away in a corner and had no qualms about making the coffee himself and offering it to every visitor with lots of food for thought on the evolving global security scenario. His columns reflected on these matters and he was also happy to answer questions from laymen on issues related to national security.

However, his stint at the newspaper did not last very long as he soon immersed himself in the task of setting up the Centre for Air Power Studies (CAPS) under a private, non-profit trust registered in New Delhi called the Forum for National Security Studies. This was mandated to undertake studies in national defence and security, military affairs and strategy in general and on air power dimensions in particular. It was meant to provide professional military education to the IAF and to help it understand the global trends in air power transformation. The first office of CAPS was a room and a half in a DDA flat, the other half being occupied by the Indian Pugwash Society. The drawing room of the flat had been converted into a small conference hall with a large round table in the centre. This also doubled up as office space once a couple of air force officers were deputed by the IAF to the Centre to initiate studies on air power and space. Most of the money for furnishing the office and hosting the few early seminars followed by lunch came out of the Air Commodore's own pocket. But, such was his passion and drive that he kept the endeavour going till more funds were provided by the Ministry of Defence to set up a corpus to sustain the Centre.

The other initial source of income for the Centre was a project that Jasjit obtained from the Department of Atomic Energy (DAE) to study the case of nuclear energy for India. Back in 2003, one day, Air Commodore called me in Jamnagar to say that CAPS had obtained the project and that it was time for some hard work. His objective was to come up with a study that would suggest solutions on how to get India out of the stranglehold of the non-proliferation regime. The task was to produce a series of papers on the Nuclear Suppliers Group – its functioning, limitations and possibilities of engagement with India. As part of this effort, a national seminar was organized in the first half of 2004 in which presentations with the idea of seeking India's exceptionalisation from the NSG were made.

As one of the deliverables of the DAE project, a newsletter called *Nuclear Power* was started. It began as a short eight-page fortnightly compilation of news and views. Despite all constraints of resources and manpower, Jasjit Singh was committed to bringing out the newsletter without interruption and to sending it gratis to policy makers and the members of the strategic community. Sitting in Jamnagar, with often unreliable access to the internet that is typical of a cantonment, I would compile nuclear news and views of every fortnight and send it to Commodore Singh at Delhi, where he had someone format it into the newsletter, make photocopies and have them dispatched. Further, during the time when the Indo–US nuclear deal was being negotiated, special issues of the newsletter were also sent to Indian missions abroad and to many members of Parliament. These endeavours would have made a difference, however small, in shaping opinions on the matter at a time when India faced a fractious debate on the subject. There were several Indian voices that objected to India accepting any safeguards on its nuclear programme as a condition of engagement into international nuclear commerce. There were also many who questioned the very need for arriving at an agreement with the USA! Amid these extreme positions, Jasjit Singh endeavoured to inject the voice of reason based on research and analyses of India's situation and global developments. He surely played a role in shaping India's nuclear narrative, but to him personally it mattered little whether his actions were having an impact or not. He just believed in doing his *karma* – actions that arose from a strong power of conviction that he derived from a compass that always pointed to national security. Unconcerned about whether his writings were being read or not, his motto was '*lage raho Munna Bhai*'!

In 2007, the scope of the DAE project granted to CAPS was expanded. In keeping with this, the newsletter *Nuclear Power* grew into *Nuclear Security*. It now covered the entire range of nuclear issues from nuclear power to nuclear strategy, non-proliferation and arms control and even missile defence. The newsletter has since been printed without missing a single issue. Jasjit Singh used to call the process of compilation of the newsletter as *akhand paath* – a continuous, unending task. It is distributed to over four hundred people today.

Over the next five years, CAPS grew in strength and stature. Several air force officers joined the Centre, undertook many projects and several books were published. Another of Jasjit's forward looking initiatives for disseminating education on nuclear issues was the idea of running week-long capsules where the officers were exposed to the best speakers on the subject. A nuclear strategy capsule was started for the senior echelons of the three services, including officers posted at the Strategic Forces Command (SFC) to explain the nuances of nuclear doctrine and strategy of India and others. From the time of the first capsule in 2007 to the twenty-sixth capsule held in January 2019, more than fifteen hundred officers have been exposed to nuclear thinking.

Having the foresight to envision the role of India in reshaping the global security order of the future, Jasjit Singh exhorted the young and not so young officers at every forum where he interacted with them to read and write more and broaden their horizons beyond their immediate professional requirements. He strongly believed that only scholar warriors could be trusted with the future of the country, especially in times of strategic uncertainty. He was concerned about defence planning for India in conditions of rapid technological change and placed emphasis on self-reliance, creation of operational capabilities and use of military power for political purposes without leading to war. His dedication, commitment and boundless energy are as well remembered as his professional experience, depth of knowledge, vision and sincerity of effort. CAPS continues down the path that he set for the organisation, with the same sincerity of purpose. CAPS, meanwhile, has grown into a well-respected national security think tank. Over eighteen years of its existence today, it has published more than ninety-two books, most of which have emerged from the research projects undertaken by the scholar faculty at the Centre.

Recognition of national security as a discipline of study

Jasjit Singh was ever ready to engage with the uniformed or the non-uniformed, the informed and the uninformed, the intelligentsia, the bureaucracy, the academia, and most of all with young students on all issues of national security. He found this imperative. In fact, his constant worry was that India was not investing enough in building 'intellectual capacity' to sustain its rise to power. Ever an optimist, he believed that India would inevitably rise to power by the sheer size of its economy and human resource potential. A greater concern for him was the need to sensitize his compatriots to the concepts and challenges of national security that would confront the country as it grew in capability and stature. Therefore, he encouraged new ideas from fresh young minds and his room was open to all. He would often say that a think tank must remain ahead of the security challenges in its thinking if it is not to become a 'thought tank'.

He was quite unhappy about the fact that the subject of national security was not taught as a separate discipline in our education system – neither in the civilian universities, military training institutions or in mass communication institutes. He considered this a major handicap in making India a knowledge society that could transcend the parochial regional affinities. He rightly believed that a country with the kind of diversity that India enjoyed must make a conscious effort to instil a 'national security consciousness' among all sections of society.

Convinced by his arguments in this matter, Kapil Sibal, Minister of Human Resource Development in late 2000s appointed an Expert Committee to examine the issues. Jasjit Singh was chosen as the head of the Committee, with Chairman of University Grants Commission (UGC) and other senior professors as its members. The report of the Committee was produced in record time and its recommendations were accepted by the Government for phased implementation. Universities were mandated to open departments of National Security Studies and the Central University of Jammu took the lead by becoming the first university to do so.

Subsequently, as member of the Indian Council of Social Science Research (ICSSR), Jasjit Singh was also instrumental in the inclusion of the subject of National Security in the list of topics on which the institution was granting research fellowships and projects to scholars. I became the first recipient of this fellowship when my proposal on space security was accepted by the ICSSR for a research grant. Needless to say, it was Jasjit Singh's idea that I make this proposal in order to get the ICSSR endeavour started off. As is evident, he was making things happen at all ends to ensure the success of his multifarious efforts at suffusing a larger understanding of the many complex dimensions of national security.

Meanwhile, as his own contribution to this process, he wrote a book titled *India's Security in a Turbulent World* which was published by National Book Trust of India in 2013. Written in an easy style for all to understand, the book was particularly meant for the young generation to make sense of the developments occurring around the world that were likely to impact India's security. In the last chapter of the book, he outlined a security strategy for the twenty-first century. He foresaw the rivalry between US and China and recommended that India 'maintain friendly cooperative relations with both the current superpower and the aspiring superpower, the People's Republic of China' (Singh, 2013:164). This had to be complemented with the build-up of military capability premised on self-reliance and with instruments to carry out effects based outcomes. The relevance of this advice in today's times is immense.

Conclusion

After spending thirty-one years in active service in the IAF, Jasjit Singh served the nation for roughly another thirty years through the study of national security and defence. Whether from the cockpit of a fighter jet or as a strategic thinker, he was consistent in his endeavour to defend, propagate and promote India's national interests in national and international military and civil circles. Indeed, he maintained national interest as the only prism through which to analyse every development in the region and beyond. And his rare ability to connect the dots to get the larger picture from tell-tale tactical details and facts helped him to peer ahead to notice what was coming. The man never let the nation down, whether as an air warrior or a strategist. In fact, he was ever ready to voice his views irrespective of how the wind was blowing and always remained practical in his approach.

Jasjit Singh was quite prolific as a writer too. He has been author and contributing editor of nearly three dozen books including such pioneering works as *Air Power in Modern Warfare* (1985), *Non-provocative Defence: The Search for Equal Security* (1989), *Nuclear India* (1998), *India's Defence Spending* (2000), *Air Power and Joint Operations* (2003), *Military Strategic Dimensions of Iraq War* (2004), *Defence from the Skies: Indian Air Force through 75 years* (2007) and *The Icon* (2009). Amongst his many affiliations, he has had some major roles in which he rendered great service to the nation. He was convener of the Task Force to set up the National Security Council (1998). He was a member of the National Security Advisory Board in 1990-91 and in 1999-2001, during which he was responsible for the preparation of India's Draft Nuclear Doctrine. Further, he served many additional roles – Member of the International Commission for a New Asia, Member of the International Commission for Peace and Food, Consultant to the Standing Committee on Defence of the Indian Parliament, Advisor as well as Member of Advisory Committee (on Defence matters) to the Eleventh Finance Commission of India, Chairman of

Committee of Experts to review the functioning of Defence and Strategic Studies and related university system of Ministry of HRD (2011), Member of Governing Council ICWA as well as of ICSSR, Member of Planning Commission Advisory Committee on Vision 2020, Member of Advisory Committee on Civil Aviation of Punjab Government, Member of Executive Council of Central Jammu University.

On Jasjit Singh's seventy-fifth birthday, K Subrahmanyam, his mentor and friend, gifted him a watch. Explaining his choice of gift, he said that since Jasjit Singh had kept a 'constant watch over India's security for the last many decades', he deserved just such a gift. And indeed Jasjit Singh had kept an eagle eye on the security of the nation. His deep insights borne out of his rich reading of history and sharp analytical acumen stood the country in good stead in several moments of crisis. There is little doubt that he stands out amongst many in the Indian strategic community as much for his balanced analysis as for his ethics and integrity to the profession.

The Padma Bhushan citation awarded to Jasjit Singh in 2006 recognized him as the country's 'leading thinker, defence expert and institution builder'. His experience, sense of responsibility, immense knowledge and ability to work hard ignoring all personal ailments was awe inspiring and worthy of emulation. For the future generations of India, here is a man who breathed his country's interests to the very last breath and passed away with a number of ideas still bubbling inside him. On his seventy-ninth birthday, on 18 July 2013, he had delivered an inspiring address at CAPS on the issue of national security – that was, of course, the only way he knew to celebrate any day! In fact, all days were equally special, or inconsequential to him. All that he looked forward to every single day was a good discussion on national security. Anyway, on his birthday in 2013, he laid down a ten-year plan for all that he wanted to achieve over the next decade. Willing to work himself to a punishing schedule, he travelled to MHOW for a seminar in the last week of July 2013, and despite taking an early morning return flight and suffering from a bad throat, he stayed on in office till late into the evening for some meetings. The bronchial infection only worsened over the weekend and he was admitted to the hospital for treatment. A week down the line, on 4 August 2013, he breathed his last. Then Prime Minister Manmohan Singh, in his condolence message, described him as a 'brave soldier', an 'inspirational military leader' who provided 'intellectual vision for India's defence and strategic planning'. Indeed, every word of that is true.

Air Cmde Jasjit Singh was a drummer who played to his own beat and marched as he thought fit. Obviously, not all appreciated this. But even his worst critics have never grudged him his integrity, forthrightness and clarity of views. Let me conclude with a short verse that I wrote for him some years ago capturing the many aspects that I, and many other scholars, have learnt from him.

> Clarity of thought and
> precision with word,
> depth of a worm and
> breadth of a bird.

Common sense and logic
over verbiage and pretence
are what you need the most
to wade from 'fog' into sense.

Being open to all –
the young and the old,
encouraging all to think,
beyond what was told.

These are only a few of the things,
that the Thinker taught all,
keeping India ahead of the curve,
an eye on every ball.

1. Jasjit Singh 2. Indira Gandhi and Jasjit Singh (far right)
3. Jasjit Singh and Arundhati Ghose 4. Hamid Ansari and Jasjit Singh

C RAJA MOHAN
An IR Scholar with a Nuclear Physics Background

Renjie Feng

IT IS WIDELY acknowledged that C Raja Mohan is one of the most remarkable foreign policy strategists and nuclear policy personalities in contemporary India. For those attempting to grasp what a nuclear India looks like today, how it has shaped up in the last three decades, how nuclear matters impact foreign policy and more, perusing Raja Mohan's works, among others, would be a good place to start. By understanding his thinking on India's security and nuclear policies through his research, writings and teachings, the interested audience has a chance to continuously remain updated on current events in the field, and witness the shaping of India's nuclear strategy of the twenty-first century.

Shouldering multiple roles – a journalist for media and press, a strategic analyst for think tanks and a professor of International Relations – Raja Mohan has his presence in many leading institutions. In India, he is associated with institutions ranging from Observer Research Foundation, Centre for Policy Research, Jawaharlal Nehru University, etcetera as well as with national dailies such as *The Indian Express, The Hindu*, and others. He is equally involved with organisations abroad where his views and lessons are considered intellectually valuable – the S Rajaratnam School of International Studies (Singapore), Nanyang Technological University and Library of Congress in Washington to name a few. Amongst the multiple accolades that Raja Mohan holds, he is best known for his expertise on foreign policy and international relations (IR). The interesting fact is he did not set out to be in the field he is now known for. Instead, he started out with a master's degree in nuclear physics. Well, the story is an important one and is narrated in the following pages.

As a journalist, Raja Mohan stands out amongst many others for his profound theoretical base in IR which allows him to view most of the current affair events from an academic prism, rather than the fast-food-style of collecting descriptive facts. As a professor, he takes upon roles that go beyond the usual responsibilities of college faculty members, who conduct lectures and often tend to limit themselves to long-term theorization or academic research. Raja Mohan with his close involvement with policy makers caters to the effective policy needs of the country. As a strategic thinker, unlike most in India who have an educational background in subjects like IR, political science, history, and other areas of humanities, Raja Mohan has a Master's degree in nuclear physics, which allows him to better understand strategic issues that involve technical knowledge about matters of nuclear science and others. Furthermore, his experience as a member of the National Security Advisory Board from 1998-2000 and 2004-06 has made his contemplation on Foreign Policy more akin to the frontline of policy formulation.

Portrayed as a pragmatist and a liberalist, Raja Mohan departs from the popular mainstream thinking on Indian Foreign Policy, which is traditionally characterized as being filled with moral aspirations, layered with ideological orientation, carrying scepticism towards the former-colonizers of the West, showing traits of antagonism towards 'Chinese aggressors' and fuelled by suspicion for Pakistan. He advocates that India look beyond the subcontinent and establish closer ties with key global powers in parallel, such as the United States, Russia and China. At the same time, he argues that India should look at conducting international relations from the prism of economics rather than that of security alone. Particularly in the case of Pakistan, Raja Mohan suggests that India use greater economic linkages to improve New Delhi's ties with its troublesome neighbour. This thinking strand has been reflected in his book *Crossing the Rubicon: The Shaping of India's New Foreign Policy* (2003) and developed in his work titled *Modi's World: Expanding India's Sphere of Influence* (2015). Consequently, his thinking on Indian foreign policy inevitably has had an impact on his reflections on India's nuclear policy, the theme we focus on in this biography.

In terms of academic and research achievements, Raja Mohan has published a host of papers in a variety of prestigious journals, which will be cited and discussed in the following sections. Besides, to date, four books based on his insightful observations and research have been published. Apart from the two mentioned above, the other two are *Impossible Allies: Nuclear India, United States and the Global Order* (2006) and *Samudra Manthan: Sino-Indian Rivalry in the Indo-Pacific* (2012). He has indeed covered a wide spectrum of topics in his writings. However, the predominant effort here would be to focus on the theme of nuclear issue, a subject that he has heavily contributed to. Starting from the 1970s, Raja Mohan worked vigorously on explicating the idea of what nuclear India is and should be. He discussed topics that ranged from nuclear deterrence, nuclear free zones, nuclear cartelization and the Intermediate-Range Nuclear Forces (INF). Moreover, he also elaborated on more recent issues like the US–Pakistan nuclear deal negotiations, US–Iran civil nuclear deal and so on, through the lenses of phenomenon description, theoretical analysis and prediction. S Jaishankar, as India's Foreign Secretary, had once remarked, 'Raja has been the predictor and the dissector of changes (of Indian foreign policy)'.[15] Certainly, Raja Mohan's various works have had an impact not only on the public's understanding of India's nuclear policy, but also on the decision-making establishment's perceptions as well as their policy options.

15 Foreign secretary Subrahmanyam Jaishankar's speech in the function of launch of Dr C Raja Mohan's book *Modi's World*, 17 Jul 2015, New Delhi. https://www.youtube.com/watch?v=_xaOpGxnVvk

This intellectual biography will categorize the evolvement of Raja Mohan's nuclear thinking since the 1970s. Efforts have been made to capture the primary characteristics of his thoughts in different periods. Firstly, from the 1970s to 1991, in the backdrop of the Cold War and the establishment of the non-proliferation of nuclear weapons regime, Raja Mohan used his pen to lash out at the 'discriminatory' notions of the regime, defying the arms control initiative from the theoretical and practical perspective. When the Cold War came to an end in 1991, Raja Mohan witnessed the intrinsic transformation of the world. Taking geopolitics into consideration, his stance began to shift towards advocating the 'great leap forward approach' that lead to the 1998 nuclear test. He enthusiastically espoused the nuclear tests conducted by India at Pokhran. He believed that it would provide India more leverages that could be used for bargaining with the international community. His views aligned with the realist point of view. This was markedly the second phase of his nuclear thinking. During the third phase, from 1998 to 2005, his nuclear thinking gravitated towards promoting a flexible response to international arms control initiatives, arguing that India abandon the previous rigid and stubborn attitude towards the CTBT (Comprehensive Nuclear-Test-Ban Treaty). This was followed by a fourth phase which came as a result of the US–India civil nuclear deal in 2005. Raja Mohan vigorously advocated that India enhance US–India relationship, and explore the relevance of the notion 'natural partnership'. After having described these various phases of C Raja Mohan's ideas on nuclear issues and its impact on shaping a nuclear India, the conclusion of this chapter notes that his pragmatism privileges contexts and circumstances over fixed norms and principles. As a thinker and scholar, Raja Mohan strives to balance relations between economics and security matters.

Phases of Raja Mohan's Nuclear Thinking

From 1970s to 1991: Defiance of non-proliferation regimes

As already noted, Raja Mohan's thinking has evolved over the years. There has invariably been a reflection of the changing circumstances of conducting nuclear politics in the world at large, and India in particular. One of the issues very often discussed by Raja Mohan has been the politics behind nuclear cartelization. In simple words, nuclear cartelization refers to a situation where countries that have nuclear technology and/or nuclear material and resources set up a cartel to prevent other non-nuclear states from acquiring nuclear technology and resources such as uranium, etcetera. (Triggs, 1976:180). This situation occurred when the nuclear haves placed uniform safeguards on all nuclear exports and only allowed the members of Non-Proliferation Treaty (NPT) to receive nuclear technology and material. This meant that the countries that did not join the NPT did not have access to nuclear technology and material, at least legally. The nuclear cartelisation was designed to force the nuclear have-nots into non-proliferation so that the nuclear haves' oligopoly and monopoly could be maintained (Coe and Vayman, 2015:984).

In the 1980s and 90s, when nuclear cartelization was being

enforced, Raja Mohan unveiled some intrinsic problems of nuclear cartelization from three dimensions. Firstly, although western nations intended to avoid proliferation through cartelization, there was rivalry and disunity among the supplier nations. Secondly, the effectiveness of cartelization was not the same as that of the OPEC (Organization of the Petroleum Exporting Countries), since nuclear material and technology are not a lifeblood as oil. Thirdly, further technological advances would defeat this cartelization as well (Jaishankar and Raja Mohan, 1977).

Another issue over which Raja Mohan expressed his indignation was Nuclear-Weapon-Free Zones (NWFZ). The establishment of NWFZ has been advocated as an approach to promote nuclear non-proliferation and disarmament norms at the regional level. Article VII of the NPT encourages states to establish NWFZ. In addition, a NWFZ is defined in the United Nations General Assembly Resolution 3472 (XXX) B as a zone of 'total absence of nuclear weapons', which is perceived as efforts to 'assist the international community in preparing for a world free of nuclear weapons'.[16] However, regarding the notion of NWFZ, Raja Mohan criticized that it can neither address the nuclear race, nor enhance international security. They can be verified neither against the nuclear power states nor the regional powers. Proposals for the free zones were flawed and more importantly, it served as a tool of

16 This information is provided by the UN Office for Disarmament Affairs, https://www.un.org/disarmament/wmd/nuclear/nwfz/

manipulation by the superpowers. He opined that it was naive to pursue 'artificial' and 'ineffective' nuclear-free zones because of pertinent reasons such as the advances in nuclear weapon technology, the drift to nuclear war fighting, arms race and so on. He warned that India should not be misled into such collaborationism, highlighting that New Delhi should stick to orthodoxy. He regarded NWFZ as a ploy to force India into an unjust nuclear world order (Raja Mohan, 1986).

He served as a member of the first National Security Advisory Board during 1998-2000 under the leadership of K Subrahmanyam, which gave him an opportunity to directly influence the government's nuclear policy. He was able to participate in the formulation of India's Draft Nuclear Doctrine (DND), on which was based the official doctrine released in 2003 (Prime Minister's Office, 2003). The DND's support for India's building and maintaining a credible minimum deterrent reflects Mohan's disapproval of the notion of NWFZ in his works.

During the Cold War, India problematized the nuclear deterrence theory that argued it could prevent war between the nuclear states due to fear of the immense destructive power of nuclear weapons. New Delhi argued that nuclear weapons, instead of preventing war, carried the potential to lead the world closer to a nuclear holocaust. At the UN Disarmament Commission in May 1965, India called for progress to be made towards 'disarmament, a comprehensive test ban treaty, a freeze on the production of nuclear weapons and the means of delivery, as well as a substantial reduction in existing stocks'

(Kerttunen, 2009: 155). India called for a treaty which could freeze and roll back the nuclear weapons that already existed, but its demands were not incorporated into the treaty. Hence, Prime Minister Indira Gandhi kept the nuclear option open, which drove her to oppose the NPT, and prepare for peaceful nuclear explosions (PNEs) (Kerttunen, 2009: 156).

Raja Mohan argued that the theory of deterrence was not only inefficient but also dangerous. The deterrence would only work if three requirements were met, i.e. capability, credibility, and communication. But the real situation didn't meet these three requirements. This could lead to the failure of deterrence. Additionally, the credibility of massive retaliation faced challenges. Technological changes reduced the stability of deterrence. US was inclined towards a counterforce doctrine and McNamara's idea of limited war in which the weapons used, the territory involved, and the goal pursued are meant to be restricted in nature. The aim of limited war is to not completely and totally destroy the enemy country. Therefore, the use of nuclear weapons is supposed to be ruled out. However, no one could be sure that limited war would not put the world on the brink of a nuclear war.

In 1982, US President Ronald Reagan appeared to prepare the US for a nuclear war with the Soviet Union, planning to spend 222 billion dollars in six years for weapons build-up and to build 17000 new nuclear weapons in the next decade (The Center for Defense Information, 1982). Raja Mohan criticized Reagan's nuclear policy that was disposed towards being a war-winning doctrine. Thus, the hope that deterrence could prevent a nuclear war was in shambles. In his paper, 'The Tragedy of Nuclear Deterrence', he pointed out that the only alternative to deterrence was nuclear abolition. However, Raja Mohan (1986: 13) pointed out that Gorbachev's 'calling of elimination of all nuclear weapons has fallen on deaf ears in the west'. From his selection of words, it appeared that the author's stance was more anti-US and pro-Russia during the 1980s. For instance, he argued that 'Soviet Union's capacity to inflict punishment on the US was less than marginal' and 'Soviet was in no position to undertake an adventurous course in Europe' (Ibid: 7). Furthermore, he applied 'the lure of counterforce' to describe the US as a war-prone power (Ibid: 16). He lambasted the aspiration of President Reagan to acquire the ability to actually execute a nuclear war. He flayed that 'for US, nuclear war becomes thinkable...for Reagan, victory is possible' (Ibid: 18).

In the last part of the same article, 'The Tragedy of Nuclear Deterrence', the negative descriptions of the 'mind-numbing strategy' of Colin Gray and the 'addiction' of US to nuclear deterrence clearly revealed Raja Mohan's antagonism towards the US. Last but not least, regarding the nuclear test ban negotiation, he pilloried that US continued to test its weapons and refused to enter into the negotiation on the Comprehensive Test Ban Treaty (CTBT). This was made clear in his article 'Nuclear Test Ban: Receding Hopes' published in 1987. It cited the following untenable reasons of the US for refusing to sign the CTBT:

- Soviets are cheats; they can't be trusted.
- Comprehensive test ban treaty is not verifiable.

- Testing is necessary to ensure the reliability of US nuclear weapons.
- CTBT would lead to the proliferation of nuclear weapons.[17]

Raja Mohan (1987) refused all the above reasons. Instead, he claimed that the real reason was that US needed testing for 'modernization' of the American nuclear power, i.e. third generation weapons. US needed more tests for investigation of the new concepts in the design of nuclear weapons and enhancement of its nuclear war-fighting capability. At the time of the Cold War when India (*de facto*) allied with Soviet Union for countering the US–Pakistan–China axis, Raja Mohan's stance during that period inevitably favoured the Soviets.

Many of his contemporaries also held a similar pro-Soviet stance. For instance, Harish Kapur (1986) lauded Soviet's diplomatic success over the US at the Reykjavik Summit held in 1986. He listed several reasons to support Soviet's stance. These included the profound concession made by Moscow to push forward the withdrawal of missiles and reduction of strategic weapons, the concrete proposals presented by the Soviets on arms control issues, etcetera. Kapur enthusiastically praised the efforts that Soviet Union had made for nuclear test ban, highlighting that Soviet Union unilaterally ceased testing of its nuclear weapons for eighteen months (Raja Mohan, 1987).

17 Raja Mohan, 'Nuclear test Ban: Receding Hopes', *Economic and Political Weekly*, vol.22 No. 7, 1987, Feb 14, p 268-270.

By the end of the 1980s, the Soviet–US tension, to some extent, had reduced. Thus the INF treaty (Intermediate-Range Nuclear Forces Treaty) was signed by the two superpowers which aimed to eliminate all land-based intermediate-range nuclear missiles and short-range missiles. At this transitional moment of international politics, Raja Mohan incisively pointed out the direction that Indian Foreign Policy should take, that India had the potential to play a role in deepening some of the positive trends emerging on the international scene.

Raja Mohan opined that India should preserve the goal of zero in nuclear arms limitation. He believed and advocated that India should urge the international community to prevent the circumvention of the possible accords on nuclear reduction through new deployments. He also suggested that the objective of India's initiative on naval nuclear arms limitation should be both reduction and rolling back of the nuclear navies towards their home territories. Further, India should press UK, France and China to join the disarmament process. He felt that India needed to assess the idea of 'comprehensive security' in order to avoid being misled. India should push for non-intervention as the central norm of world politics and should explore and shape the debate on the role of the UN (Raja Mohan, 1987b). With these, he made his clear case for the country's policy options in context of nuclear issues. Raja Mohan underscored that India should not simply dismiss the 'peace proposals' by the Soviet and other countries for arms limitation (Raja Mohan, 1987b:4). According to him (1987b: 23), although the arms reduction treaties such as the 1987 Intermediate-

Range Nuclear Forces Treaty (INF) has been one of the most significant progresses in arms control, India should contend that it was not enough to safeguard its interest in the best way. Rather, India should 'devise and diplomatically project a set of peace proposals encompassing both nuclear and broader security issues, including the criteria for peace' such as 'non-discriminatory nature, opposition to foreign presences, de-escalation of military tensions' (Ibid: 23).

From the articles cited and discussed above, the characteristics of Raja Mohan's thinking from 1970s to 1991 (before the collapse of Soviet) can be drawn. He influenced India's views on nuclear issues, and subtly, if not directly, shaped the policy decisions of the establishment at different phases of India's nuclear diplomacy. He was a persistent defier of international non-proliferation and arms control regimes, critically assessing topics such as nuclear cartelisation, nuclear free zones, and deterrence theory, which serves as the underpinning or the supplement of the NPT. While he denounced the discriminatory nature and the failure of the NPT regime constructed by the haves, he was still more inclined to view Soviet as a more benign and peace-loving country than the US, a country crucified by him as the creator of a host of international problems, viz. arms race, moving close to all-out nuclear war, etc. In any case, through the works and lens of Raja Mohan's research, there is a lot to learn about the various themes on nuclear power politics at the international level.

From 1991 to 1998: Proponent of nuclear great leap forward

Since the collapse of the Soviet Union in 1991, the world has witnessed an intrinsic transformation, moving from bipolarity to multipolarity. Undoubtedly, the end of the Cold War has had a profound impact on the world as a whole and on the states as individuals. However, the implication for India was unique due to a host of reasons. For instance, it is the only democratic state that had almost allied with the Soviet Union. Also, India faced threats from two neighbouring countries which were members of the US-led 'axis'. Under these circumstances, India faced huge uncertainties about its security as the protection from the Soviet Union waned. The incentives to go for nuclear weapons increased dramatically. Even the criticisms of such a 'leap forward', from a rather moral perspective, and the fear of severe sanctions against India, which could undermine its economy, did not seem as robust reasons to restrain itself from testing nuclear weapons.

Raja Mohan argued that it was such traditional inhibitions that were dragging India's leg to 'cross the rubicon'. Both Indira Gandhi and Narasimha Rao pulled back when they had the opportunity. Though the former gave the green signal for Pokhran-1, India held back from announcing itself as a nuclear weapon state in 1974. India did not risk shedding the image of being a morally responsible power that only advocated the peaceful use of nuclear power. This hesitance on New Delhi's part is looked upon with regret by many scholars and strategic experts, including Raja Mohan.

Raja Mohan expressed that it was the traditional inhibitions that left India confronting with most of the international non-proliferation orders (Raja Mohan, 2004).[18] In other words, Raja Mohan implicitly meant that India would have sat on the high table if India had overcome its moral concerns and conducted nuclear tests earlier. He believed it was high time India seized the opportunity in 1990s to put its nuclear ambiguity to an end, and to declare its nuclear status. Such an urgent appeal is reflected when he opined, 'if India could not make up its mind on what to do with nuclear weapons, the world would decide for it by demanding India give up its nuclear option' (Ibid., p.3). He wrote, 'once again all the traditional inhibitions that prevented it from taking the nuclear leap seemed to come back into play'. Traditional inhibitions and moral concerns have not interested Raja Mohan. Instead, he advocated a policy of 'the earlier, the better' in terms of nuclear weapon tests by India. The BJP government eventually decided to go for the testing in 1998 after three attempts that suffered setbacks. Further, Raja Mohan lauded the United Front government for their right decision to bear the near-term political and economic costs for obtaining the long-term gain.[19] Raja Mohan's predictions, his advice and his foresight of India's nuclear scenario are qualities that beget robust strategic policies for future development and progress in India's nuclear journey.

---

18 Raja Mohan, *Crossing the Rubicon: The Shaping of India's New Foreign Policy*, New York: Palgrave, 2004, p. 6.

19 Ibid, 14.

From 1998 to 2005: Transformation from engaging with US to cooperating with US

After Pokhran-II, unsurprisingly, India's relations with the sole superpower, US, became worse. It resulted in a greater cost for India at many levels, particularly due to the economic sanctions levied on the country. However, from Raja Mohan's viewpoint, the biggest cost was that India demonstrated its nuclear potential to Pakistan, in turn opening up a path for Islamabad to follow suit. Due to the threat of severe sanctions on its already fragile economy, it would have been difficult for Pakistan to make the first move in terms of testing nuclear devices, if India had not done so in 1998. Had Pakistan decided to go nuclear first, the cost would have been unbearably high for the Nawaz Sharif administration. On the other hand, it would have placed India in a comparatively comfortable position to justify New Delhi's decision to go ahead with *Operation Shakti*, i.e. present it as a response to Islamabad's testing of thermonuclear devices. In any case, India's declaration of its nuclear status provided Pakistan with a strong reason for which it had waited long. The nuclear test explosions in Pokhran in May 1998 put Pakistan in a better position to justify its own tests a few weeks later in the guise of responding to India's nuclear blackmailing. Consequently, Pakistan went ahead with six underground nuclear explosions of varying intensity at Chagai in Balochistan. By doing so, Pakistan overcame the disparity in terms of conventional weapons and troops with India.

Pakistan's leaders believed that India would treat Pakistan more seriously and equally after the ambiguity about the two countries' nuclear capability was removed. As Pakistani scholar Mahmood points out, 'It was in the best interest of Pakistan to explode its nuclear device to be at par with India (1999: 39).' Raja Mohan, much like many others, was convinced that the reasons for India's 1998 nuclear tests were rooted in the deteriorating security environment after the Cold War.[20] For instance, India perceived a 'China threat', especially without the protection of the Soviet's nuclear umbrella. The nuclear parity between the Soviet and US was not in place anymore in the post-Cold War era. Hence, India had to consider declaring its own nuclear capability for its security.

As suggested by Raja Mohan, after the Pokhran-II tests in May 1998, there were two ways to overcome the challenges posed by the global nuclear order. One was dealing with India's economic and commercial policies. The other one was to negotiate a modus vivendi with the global nuclear order. If the great powers had refused to recognize India's nuclear status, India had two options:

- Negative and multilateral: India would have to step up its rhetoric on disarmament, challenge the nuclear monopoly of the P-5, question their nuclear doctrine, reject all interim arms control treaties. India's proposal of universal disarmament not only appealed to the international community and the UNSC, but it also found support at the domestic level.
- Positive and unilateral: India could demonstrate that it would

20 Vajpayee's letter to the US President was leaked in the NYT.

be a responsible nuclear weapon power, emphasizing the limited role of nuke, minimum deterrence and un-provocative nuclear doctrine.

In other words, he argued that the world was left with only two choices – to either integrate India into the global nuclear order, or provoke it to take a rather extreme position.[21] It reflects Raja Mohan's disapproval of the constraints and discrimination that the US-led non-proliferation community imposed on India. After India detonated a nuclear warhead, he shifted from strong aversion to a more pragmatic stance, providing two options for India's decision makers – a flexible and practical choice and a stubborn and resistant stance – depending on how the international community would treat India.

However, certain shifts in India's nuclear trajectory began to occur with the turn of the millennium. In March 2000, US President Bill Clinton paid a significant state visit to India, intending to shift his India policy from sanction and isolation to engagement and integration. Meanwhile, US also recognized the necessity to maintain benign ties with Pakistan, which placed India in a rather competitive mode. Although disagreements appeared within the US State Department regarding Clinton's visit to Pakistan, the US President eventually decided to have a stopover in Islamabad after his visit to India. Thus, the ongoing US–Pakistan interaction at that juncture

21 Raja Mohan, India and Nuclear Weapons, http://library.fes.de/pdf-files/ipg/ipg-1998-4/artmohan.pdf.

also drove India to take a more pragmatic approach and seek to constructively engage with the US.

After US President Clinton's visit to India, Raja Mohan's views gradually transformed to advocate pragmatism with the US. He suggested seizing the opportunity to enhance the Indo–US relationship. This finds ample space in his master work, *Crossing the Rubicon: The Shaping of India's New Foreign Policy* (2004). One of the paramount changes of Indian foreign policy, listed in the book, was the glaring shift from an anti-western mode of thinking to rejection of that mentality. His contemporaries also gave similar opinions about India's evolving relations with the US in the early 2000s. For example, Achin Vanaik (2008), who has also contributed immensely to India's nuclear narrative, argues that 'it is in India's interest then to bandwagon behind the US'. During the Clinton administration, India had also successfully convinced the US government that the US should accept India's nuclearisation as India was a strategic asset for the US. Similarly, as Rajesh Basrur (2017: 7) points out, India realized that its 1998 tests had increased the US's interest in building a strategic partnership with India. India too was willing to seize the opportunity to enhance its ties with the US, which eventually led to the India–US nuclear deal in 2005. This helped India to 'bypass the tightening rules of nuclear commerce despite its refusal to sign the Nuclear Non-Proliferation Treaty (NPT)'.

Raja Mohan opined that the nuclear tests in 1998 rendered India important leverages because of the following reasons. Firstly, it is said that India did not need to have any more nuclear tests after Pokhran-II since the data collected through the five tests of Operation Shakti were already sufficient for India's nuclear weapon programme. Therefore, it created the possibility for India to enter into the negotiation of CTBT, which could then serve as a bargaining chip. Secondly, because of India's nuclear tests, US treated India more seriously. The de facto nuclear status of India allowed India to negotiate and bargain with great powers (Raja Mohan, 2004:16). The nuclear debate mentioned above between the US and India was an effective and productive engagement, which helped to improve bilateral relations. Additionally, in the last few decades, one of the impediments between the two states was India's concept of non-alignment. Although India did not give up on the notion of non-alignment, its relevance reduced dramatically by the end of the 1990s and early 2000s, paving the way for strengthening Indo–US relations.

Subsequently, after the Indo–US relation began to normalize, it became inevitable for the US to raise the issue of integrating India into the international nuclear order. In terms of CTBT, Raja Mohan believed that it made little sense for India to continue to oppose the CTBT since it had already tested nuclear weapons. However, he contended that it would be too rapid for India to accept the treaty immediately, given that India took a strong position against it in 1995 and 1996 (Ibid: 16). What Raja Mohan advocated was that Indian decision makers should abandon or at least reduce the inertia of old thinking, which treated signing of the CTBT as a harm to its sovereignty or as submission to western pressures. Instead of embracing the previous inflexible position on CTBT, Raja Mohan suggested that the government should apply a more pragmatic approach to participate in the negotiation of CTBT, deciding whether to sign CTBT or not according to rational cost-benefit analysis (Ibid: 19).

Interestingly, writing about it, Raja Mohan contends that India and the US are not allies in the traditional sense, not even in the foreseeable future, but areas of agreement and common endeavour have begun to replace the negative agenda of the earlier wasted decades, which is a significant milestone for both. However, Raja Mohan warned that if India starts the security cooperation with the US, it would involve short-term political risks that might outweigh long-term advantages. Given that India is propelling its enterprise to build a modern capitalist society, therefore, India would define its alliance with the US in economic terms, rather than security terms. Even though there has been increasing security cooperation between the two, India seemingly prefers to direct it for defensive purpose, rather than for offensive expansion.

From 2005 onwards: Advocate of expanding the scope of Indo-US relations

Before 2005, although Raja Mohan enthusiastically advocated the enhancement of Indo–US relations in TV discussions, newspapers and IR seminars, some reservations and a certain level of scepticism about the Indo–US security cooperation was tangibly visible in his thinking. However, since the signing of the Indo–US civil nuclear deal in 2005, Raja Mohan, instead of showing reservation or scepticism on security cooperation with the US, has endorsed a more robust Indo–US bonhomie, including in security issues.

Raja Mohan, tracing the evolution of the bilateral relationship, said that the idea of a strategic relationship between India and the United States 'could not have been visualized' for much of the Cold War era. During that period, the relations were abysmal and dominated by ugly stereotypes about the 'insolent Indian' and 'ugly American'. The situation began to change in the early 1990s, when India initiated economic reforms, marking a shift in its economic orientation and creating the basis for a new type of relationship with the United States. Additionally, the Indian-American community – which now numbers well over a million – has become a 'bridge' linking the two countries. This was highlighted during Prime Minister Narendra Modi's visit to the US in September 2019 and the astounding event of 'Howdy Modi' organized in Houston.

Yet, before that, Raja Mohan showed concerns about the relationship between India and US during the Obama years, worrying that the bilateral ties may go down as the Obama administration could possibly come closer to Pakistan and China. To some extent, Raja Mohan's concern was not without reason. The lack of focus on India in the first six months of the new administration was illustrated by the fact that Secretary of State Hillary Clinton visited China and other countries, but not India. In addition, Richard Holbrooke, the US Special Representative for Afghanistan and Pakistan, mentioned that he might include the Kashmir dispute in his agenda, which made Indian officials feel the new administration was not respecting Indian core national interests enough. Additionally, the Obama administration was also interested in enhancing the US–Pakistan relationship and provided Pakistan $3.2 billion military assistance for the war on terror. This stood the chance of diminishing India's conventional military advantage vis-à-vis Pakistan. Last but not least, in terms of the Sino–US relations, the Obama administration also engaged actively with China in order to get Beijing's help on Afghanistan and Pakistan.

Raja Mohan categorically stated that the US could not manage global challenges alone and would need India to stand alongside it at many points.[22] Similarly, in 2003, Raja Mohan, in his speech at the Heritage Foundation, asserted that the, 'Indo–US relationship has been the fastest-growing bilateral relationship. I don't think any other bilateral relationship for both countries has expanded so much in such a short period of time.' Meanwhile, he listed impediments which could hinder both countries from moving beyond a plateau stage, including the slowing down of economy, different strategic culture, etc. (Raja Mohan, 2013a). At the end of his speech, he highlighted optimistically that since the basis of the Indo–US relationship has been anchored on many common interests, and the problems between India and US are relatively few, it is reasonable to expect the bilateral relation to move forward

22 See Raja Mohan's speech in the conference on strategic dimension of the US-India relationship, organized by the Asia Program of Wilson Center, Jan 27, 2010. https://www.youtube.com/watch?v=uWWUe2iiN2I.

if both sides remove the hurdles on the way (Ibid.). Moreover, he pointed out that emerging China also played the role of a catalyst in driving the US and India closer to each other out of the realist need for balance of power (Raja Mohan, 2013b).

In terms of the Indo–US civil nuclear deal, Raja Mohan argues that although India did not buy a nuclear powered reactor from the US, this deal nevertheless transformed the bilateral relationship that reflected a few noteworthy achievements:

- US became India's largest trading partner in goods and service.
- Cooperation began to be realized on counter-terrorism and intelligence-sharing issues.
- US removed many high-tech sanctions imposed on India since 1974.
- US refused to extend the civil nuclear initiative to Pakistan.
- The two states made efforts to avoid confrontation on trade and climate change issues.

Witnessing the progress in bilateral relations, Raja Mohan recognizes that political ambivalence and scepticism have not however diminished in the decade since 2005 (Raja Mohan, 2015a). For instance, there were quite some concerns that India's deepening alignment with the US might have an impact on India's independence in its foreign policy decisions. As Smith (2020) mentions, for most of the US allies, 'each step closer to the United States results in a corresponding loss of strategic autonomy'.

When Raja Mohan talks about India's 'nuclear exceptionalism', a term that has come to be commonly used in nuclear literature, he argues that there are two reasons accounting for such an appearance of exceptionalism. Firstly, after 1998, India seemed to have played a more responsible role in its arms control policies. India became more ready to accept internationally mandated restrictions on its strategic programmes, to recognize weapons of mass destruction (WMD) proliferation as an important international security problem, to implement controls on the spread of sensitive technologies, and to accept the need for a credible regime of nuclear and conventional confidence-building measures (CBMs) in the subcontinent to reduce the danger of a nuclear war. Thus, India, to some extent, relaxed the previous resistance and became more willing to contribute to the arms control regimes, which in turn relieved tension off India's 'refusal to sign' stance and the international nuclear order. Secondly, due to the changing nature of the Indo–US relations, the two sides chose to put in place a broader engagement framework, including to recognize a 'natural alliance' partnership and increase security cooperation (Raja Mohan, 2007:113-151).

Apart from his detailed investigation on the subject of India–US relations, particularly in context of its nuclear cooperation and interfaces, Raja Mohan also explicates India's policy options vis-à-vis Pakistan's nuclear strides. In many of his writings, he discusses the nuclear relations between India and Pakistan. He contends that the US–India deal would alter India's status in the non-proliferation regime, and end the sense of legal equivalence between India and Pakistan, though the nuclear threat from Pakistan remains. Raja Mohan, while interacting with Robert J Einhorn, a senior adviser at the Center for Strategic and International Studies in the US in 2005, described the Indo–US nuclear agreement as a 'historic' 'big deal' that did not force India to put a cap on its fissile material usage while allowing India to involve in civil nuclear deals (CSIS, 2005). Though Einhorn stated that he favoured the improvement of the bilateral relations, he expressed reservations about the bilateral nuclear deal. He hoped the Bush administration would demand more concessions from India such as putting a restraint on India's own fissile material usage and pushing India to put more pressure on Iran regarding the latter's nuclear program (CSIS, 2005). Furthermore, Raja Mohan suggests that India shed its traditional reluctance to work with the US to manage all terrorism threat emanating from Pakistan. He refutes Pakistan's justification of its attempt of seeking parity with India and advocates that India should work in tandem with the US and with the moderate forces in Pakistan. According to Raja Mohan (2008: 139), when the situation requires, India should take responsibility to pacify borderlands. Meanwhile, he also criticized the US policy of hyphenation which prohibits India from having a say in US policy towards Pakistan (Raja Mohan, 2008). His elaborations on not just India's nuclear issues, but also on other countries' nuclear issues reflect well his clear and pragmatic nuclear thinking. For instance, he also opines on US–Pakistan negotiation and US–Iran nuclear accord with practical suggestions.

In terms of the US–Pakistan nuclear negotiation, thinking beyond the simple traditional mindless denunciation of

Pakistan, he enthusiastically advocates that India should think of exploiting opportunities in the US–Pakistan nuclear negotiations. Instead of opposing the entire deal, he suggests India should try to shape the outcome in favour of its own interests by injecting issues relating to terrorism and democracy into the US debate with Pakistan. His view is based on the reality that given US's strong hyphenation policy in the past towards these two rivals of the subcontinent, it would render India little benefits if it overtly and harshly criticizes the US–Pakistan nexus. Therefore, rather than outraging alone outside of the negotiation, he advocates India should 'think with [its] head rather than gut' for influencing the negotiations. India should take its lessons from the past when it successfully negotiated benefits for itself at the time when Bush resumed arms sales to Pakistan in 2005 (Raja Mohan, 2015b). It is noteworthy that Raja Mohan's arguments have evolved from his previous stance of problematizing Pakistan's legitimacy to seek similar nuclear status to promoting the magnification of India's impacts on the process of US–Pakistan negotiation.

In terms of the US–Iran nuclear bargaining, Raja Mohan (2015) argues that the two countries' nuclear accord would render new opportunities as well as challenges for India. In terms of the advantageous dimensions, due to the accord, firstly, the relaxation of US sanctions on Iran would bring large amounts of Iranian oil into the market, which in turn will contribute to the descent of petroleum prices. It would also create new business possibilities for Indian companies to explore Iranian market. Further, India's options in Central

Asia and Afghanistan will expand rapidly, in the case that Iran serves as a gateway for India to further its engagement with the two regions. Meanwhile, Raja Mohan (2015) reminds that the challenges should never be ignored. Since the Middle East has witnessed severe tensions between Iran and Israel, Raja Mohan (2015) suggests that India needs to navigate cautiously given the region's new contradictions. This line of thought can be found in an article he penned titled 'How the Iran Nuke Deal Gives India Room in the Greater Middle East'. At the end of the article, Raja Mohan highlights the importance of bilateral coordination between India and Iran. He says that, 'if America and Iran do begin to cooperate against violent extremism in the Greater Middle East, it should have many positive consequences for combating terrorism in the subcontinent'.[23]

Raja Mohan's views, his advice and scholarly researche have provided much food for thought, as well as clear contours for action, to the strategic and nuclear community in India, among many others. Not only matters that are nuclear and security related but also various dynamic issues of foreign policy and international relations remain Raja Mohan's field of scholarly investigation. His regular columns in *The Indian Express* and its wide readership even today speaks volumes of his continuing active career that has only shone over the last decade. No doubt there are people who have possibly done more to give shape, impact and analyse India's nuclear thinking in recent times,

---

23 C Raja Mohan (2015), How the Iran nuke deal gives India room in the Greater Middle East, http://indianexpress.com/article/explained/how-the-nuclear-deal-gives-india-room-in-the-greater-middle-east.

but Raja Mohan's contribution in the field stands out for its comprehensiveness, audacity and practicality to say the least. In any case, continuous and clear foreign policy suggestions, specially pertaining to nuclear matters, coming from a person with understanding of nuclear physics is more valued and even rare in the country.

Conclusion

Raja Mohan obtained his PhD degree from Jawaharlal Nehru University in early 1980s. Since then, he has been involved in various IR-related jobs. From being a correspondent and later an editor for newspapers, a member of the Indian National Security Advisory Board, a professor at universities at home and abroad, an academic at think tanks, to a founding director of Carnegie India, among other responsibilities, Raja Mohan's career has inspired a whole generation of people. Although this chapter has attempted to exclusively focus on his contribution to Indian nuclear thinking, nuclear diplomacy is merely one of the wide spectrum of research areas that Raja Mohan has energetically covered in his career spanning many years. Such unique and vibrant career experiences with rich diversity have empowered him to look at international affairs from various prisms, which has sharpened his point of view making him extremely insightful. His expertise is well recognised by the IR circles in India and abroad. Raja Mohan was appointed Director of the Institute of South Asian Studies (ISAS) at the National University of Singapore in May 2018. Speaking on the occasion, renowned Singaporean diplomat Gopinath Pillai said, 'I strongly believe that Professor Mohan is well qualified... He comes with a wealth of experience on South Asia and its relations with the global powers.' As the director of the leading think tank in Singapore, a small island country known for its strategic location in Southeast Asia that is thought to be the centre of Indo-Pacific, Raja Mohan has propelled his career to greater heights. He served in this position for two years till December 2021. Now Raja Mohan is associated with ISAS as Visiting Research Professor. He will continue to exert his influence on international politics in the new era featured by the formulation of security complexes in the Indo-Pacific, be it the Quad, the AUKUS or any other multilateral forum.

This chapter, through the chronicling and dissection of his works on India's nuclear diplomacy, in particular from 1970s to 2015, draws a clear conclusion that pragmatism is profoundly featured in his nuclear thinking. Two parameters could be applied to explain his pragmatic thinking. Context and circumstances prevail over principles and norms. When Raja Mohan opines his stance on India's nuclear policy and other relevant foreign policy issues, he does realize the importance of standards and morality. However, providing policy advice on defending national interests and expanding the state's scope of external influence are primary goals for a professional strategist. Therefore, his suggestions and views on certain issues will depend more on the specific matrix rather than fixed principles and norms. Several examples could be introduced to demonstrate this strand of his thinking. Firstly, during the Cold War, his articles (written in late 1970s and during 1980s) harshly criticized US, who shared the same democratic values as India, and praised the arms control efforts of the Communist Soviet Union, which followed a totally different ideology. However, when the Cold War came to an end and the external security environment was fundamentally changed, his views on US evolved to advocate India's constructive engagement with the superpower. Additionally, it was perceived that it would never be favourable for India if US and Pakistan came close to each other on nuclear issues. Likewise, Mohan used to support India's exceptionalism and deny Pakistan's legitimacy to acquire the privileges granted to India in the 123 Agreement. Yet, in the post-US–India nuclear deal period (especially during the Obama Administration), he comprehensively took the international circumstance and Washington's geopolitical consideration into account, and suggested that India should seek to influence the direction and outcomes of the US–Pakistan negotiation, rather than simply and brainlessly opposing it.

Secondly, in his strategic thinking, he strives to strike a balance between economic relations and security concerns. Traditionally, Indian decision makers are preoccupied with security and political issues, neglecting the importance of regional trade on a state's long-term benefits. This does not imply that Raja Mohan doesn't place enough emphasis on security. But for him, it may not be the most feasible approach to maximize India's interests in the short term or the coming future by taking rigid and stubborn positions out of security considerations which would impede bilateral trade. For instance, the aversion towards Pakistan and scepticism towards China are widely spread and deeply rooted in India. However, as a visionary scholar, Raja Mohan advocates to institute trade with the former and enhance the economic linkage with the latter, rather than unsophisticatedly looking at these two neighbours through the narrow prism of security as was prevalent during the Obama situation.

However, this observation in the chapter only refers to the situation before 2016. Since 2017, due to India's deteriorating relations with Pakistan and China, the formulation of Quad and the emerging notion of Indo-Pacific, the situation has changed tremendously, and the discussions of security issues have significantly gained momentum. The mainstream view of the Indian strategic circle, including Mohan, has called for 'active economic decoupling from China' (Raja Mohan, 2020).

REFERENCES

- Foreign Secretary Subrahmanyam Jaishankar's speech in the launch function of Dr C Raja Mohan's book *Modi's World*, 17 July 2015, New Delhi, https://www.youtube.com/watch?v=_xaOpGxnVvk.
- S Jaishankar and C Raja Mohan, 'Nuclear Cartelisation: Theory and Practice', *Economic and Political Weekly*, Vol. 12, No. 20 (14 May 1977), p. 798-802.
- C Raja Mohan, 'Nuclear Free Zone: Illusion And Reality' in

K Subrahmanyam (ed.), *India and the Nuclear Challenge* (New Delhi: Lancer international 1986), p. 143-167.

- C Raja Mohan, 'The Tragedy of Nuclear Deterrence', *Social Scientist*, Vol. 14, No. 4 (Apr 1986), p. 3-19.
- C Raja Mohan, 'Nuclear Test Ban: Receding Hopes', *Economic And Political Weekly*, vol.22 No. 7, 14 Feb 1987, p. 268-270.
- C Raja Mohan, 'India and the Asian Security Concept', *Economic and Political Weekly*, Vol. 21, No. 47, 22 Nov 1986, p. 2037-2039.
- C Raja Mohan, 'The Peace Scenario After The INF Treaty', *India International Centre Quarterly*, vol. 14, No. 4, 1987, p. 1-23.
- C Raja Mohan, *Crossing the Rubicon: The Shaping of India's New Foreign Policy* (New York: Palgrave, 2004), p. 6.
- C Raja Mohan, 'India and Nuclear Weapons', <http://library.fes.de/pdf-files/ipg/ipg-1998-4/artmohan.pdf>.
- C Raja Mohan's speech in the conference on Strategic Dimension of the US–India Relationship, organized by the Asia Program of Wilson Center, 27 Jan 2010, <https://www.youtube.com/watch?v=uWWUe2iiN2I>.
- C Raja Mohan's talk in the Heritage Foundation, 'Future of US-India Relations: Beyond the Plateau', 2 May 2013, <https://www.youtube.com/watch?v=ot6FGUCJz78>.
- C Raja Mohan's speech in Princeton on the topic 'India and the US Pivot to Asia: Strategic Autonomy or Geopolitical Opportunity'
- C Raja Mohan, '10 Years of Indo–U.S. Civil-Nuclear Deal: Transformation of the Bilateral Relationship is the Real Big

Deal'. 20 July 2015; *Indian Express.* <http://indianexpress.com/article/explained/10-yrs-of-indo-us-civil-nuclear-deal-transformation-of-the-bilateral-relationship-is-the-real-big-deal>.

- C Raja Mohan, 'India's Nuclear Exceptionalism', in Morten Bremer, Maerli, and Sverre Lodgaard (eds), *Nuclear Proliferation and International Security*, Rutledge Global Security Series, 2007, p. 113-151.
- C Raja Mohan, The Great Nuke Game, *India International Centre Quarterly*, Vol. 35, No. 3/4, the Great Divide (2008), p. 130-140.
- C Raja Mohan, 'Raja-Mandala: Turning the Debate to Delhi's Advantage', 20 Oct 2015, <http://indianexpress.com/article/opinion/columns/turn-the-debate-to-delhis-advantage>.
- C Raja Mohan, 'How the Iran Nuke Deal Gives India Room in the Greater Middle East', 15 July 2015, <http://indianexpress.com/article/explained/how-the-nuclear-deal-gives-india-room-in-the-greater-middle-east>.

2

1. Raja Mohan with WISCOMP Director Meenakshi Gopinath
2. Raja Mohan

SECTION D

DIPLOMATS

ARUNDHATI GHOSE
The Proud Indian at the NPT Negotiations

Vijaylaxmi Yadav

Onnay je kore, Aar onnay je sohe,
Tobo ghrina jeno trino somo dohe[24]

IF EVER THERE was one, the above proverbial phrase by Rabindranath Tagore is an apt description of Ambassador Arundhati's life philosophy. She was an independent woman, an upright civil servant, and an astute diplomat. In diplomatic circles, she is often remembered as being honest to a fault, practical and courageous. With an illustrious diplomatic career spanning over three decades, Ghose is most well-known for her role in the Comprehensive Test Ban Treaty (CTBT) negotiations in the Conference on Disarmament (COD). She also served India as the liaison officer in-charge of the

24 The phrase literally, translates into – The one who does evil/wrong; and the one who tolerates the evil/wrong; May your disgust burn both of them like straw. The phrase highlights the complicity of those who stand by mutely in the face of wrongdoing/ injustice.

government of Tajuddin Ahmad during the 1971 Bangladesh war, and played a crucial role in the survival of the government in exile.

The following essay is an attempt not only to trace her professional and academic life, but also to illuminate some parts of her personal journey through life and the Foreign Services, and the various identities she donned throughout her life in a personal capacity, as remembered by those close to her. The most striking element of these recollections, however, remains her proud self-perception as an Indian – a progressive modern woman who was a patriot to the core.

The current essay attempts to reconstruct an intellectual biography of Ghose, based on her work as the Representative of India at the COD as well as an academic and influential intellectual. Wherever possible, it also includes comments from people who were associated with her in different personal and professional capacities. The chapter is structured in such a way that the selected works by Ghose are studied and summarised individually to form a large corpus, and finally, some conclusions are drawn. Her work is summarised and core arguments are presented in an attempt to place her position in the larger International Relations theoretical framework.

A personal profile

Arundhati, or Chukku, as she was fondly called, was born into a prominent Bengali family in 1940, and grew up in Bombay. She had three siblings – two older brothers and a younger sister. Her early years were spent in Bombay, and subsequently in

Delhi. The untimely death of her father, who was a government servant, pushed the family to a life of genteel poverty and triggered the move from Bombay to Delhi. She grew up in a loving and enabling home and had a childhood filled with adventures. Her brother, Bhaskar Ghose, who is also a retired bureaucrat, recounts the siblings going for long walks and treks to the Mehrauli fort through the vast expanse of forest which lay in between the fort and their Uncle's home in Sarojini Nagar (then called Vinay Nagar), Delhi, where they had moved. The early years of hardship instilled in the children, and in a young Arundhati, the determination for life, and unconditional love for and loyalty towards family. These would be attributes that continued to define her life and actions till the very end. She passed away on 25 July 2016 after a protracted battle with cancer.

Ghose had decided to become an Indian Foreign Services officer early on in life. She was a strong-willed, yet quiet and reserved youngster, who once decided on something was not easily swayed. She graduated in English from Lady Brabourne College in Calcutta. She then completed her Masters – also in English – from Visva-Bharati University in Santiniketan, West Bengal, while simultaneously taking tuitions for school children to supplement the family income. Eventually, she moved to Barrackpore to live with her mother and brother Bhaskar (who was then posted as a Sub-Divisional Officer there) and started preparing for her exams. She was hardworking and focused, and determined to serve her country in the capacity of a diplomat.

Her love for the country was matched only by the

unconditional love for her family. While she did not marry, she was greatly involved in the lives of her siblings, and subsequently their children. Amidst doubts from the family, Ghose, as a young diplomat beginning her career in Vienna, supported her younger sister Ruma while she completed a Bachelor of Civil Law (BCL) from Oxford University. It was her love for her sister, and faith in her intentions, which helped both of them get through a period of hardship and compromise. The two sisters were inseparable, and remained so till the end. Ruma Pal retired as a Supreme Court judge in July 2006.

Wherever Ambassador Ghose was posted throughout her career, her siblings and their children invariably visited her and were regaled with lots of stories, love, and warmth. Her family remembers her as a very hospitable and caring woman for whom it was instinctive to be by her family's side, through thick and thin. One such instance is recounted by Bhaskar Ghose. When their mother became seriously ill, both Arundhati and Ruma stayed by her side to care for her. The room that had been found in AIIMS (All India Institute of Medical Sciences, Delhi) through the Bharatiya Janata Party (BJP) leader AB Vajpayee's timely assistance, was scrubbed clean by both the sisters themselves to ensure their mother's admittance to the hospital. Another instance is from her time as Minister Counsellor in Belgium to the then European Economic Community. She insisted that her brother Bhaskar, who was at Oxford for a fellowship at the time and was alone for the winter break, visit and spend time with her in Brussels.

Ghose was known as a lively fun-loving person amongst

her friends: she loved to entertain, and to have a good time. Her niece, Sagarika Ghose describes her visits to her aunt (when she was posted in Brussels and Paris) as being curated with care so as to ensure she – and her friends who sometimes came along – always felt welcomed and loved. She took them out for long walks and drives, engaged them in conversation, and discussed general and specific issues of life passionately. Bhaskar Ghose also mirrors this sentiment of being warmly received, and treated with enormous love every time he visited his sister.

Personally, she was a voracious reader with favourites including Mary Stewart, Georgette Heyer and Margaret Drabble. Books were also her favourite travel souvenirs. Her book shelf today, preserved by her brother with care, has Thomas Hobbes and Agatha Christie on the same shelf – a reflection of the wide range of her personal and academic interests. She also loved listening to music and jived to Boney M. Her favourites included Joan Baez, Bob Dylan, Peter, Paul and Mary, Nana Mouskouri, Paul Anka, the Beatles, and Pete Seeger. She smoked with relish, incessantly and unapologetically. She would always wear only her trademark silk sarees, often with a sleeveless blouse, in the early years of her career, so much so that Sagarika Ghose credits her aunt for her choice of style in clothing. While she was not a connoisseur, she took great care of her sarees, and ensured that she presented the very best of her country's handlooms on the numerous occasions she was its representative at the international platform. This, however, did not prevent her from hitching up her saree to go cycling in a park near her home in Brussels when she felt like it. In all respects, from her hair to her life choices, she was an unconventional Indian woman of her time.

Arundhati Ghose – The professional, academic, and intellectual

Professionally, Arundhati Ghose is remembered as a hardworking and dedicated diplomat who loved what she did, and did it with utmost sincerity and pride. She was vehemently opposed to intellectual dishonesty and did not tolerate pretences or hypocrisy. The pride in her work stemmed from her undying love for her country and its interests, which was always her first priority. This is also why one always witnessed a very guarded and measured version of Ghose on international platforms, who was otherwise known for her infamous temper.

Arundhati Ghose also showed a great deal of loyalty to her superiors and mentors including diplomats Jyotindra Nath Dixit, Parmeshwar Narayan Haksar, and Muchkund Dubey. She is also known to have admired and worked closely with Atal Bihari Vajpayee, the then Foreign Minister and LK Advani. During her period of service, which started in 1963, she was posted to Missions in Austria, the Netherlands, Belgium (Brussels for the then EC), Bangladesh and New York (in the Permanent Mission to the UN) and as Ambassador to the Republic of Korea, Egypt, Permanent Representative to UNESCO in Paris, the UN Offices in Geneva, and Ambassador to the Conference on Disarmament, Geneva.

In her statement explaining the vote in the COD on

September 1996,[25] Ghose focused on the discrepancies between the resolutions passed in the United Nations General Assembly (UNGA) that day and the one in whose adoption India had participated in 1995. The central issue was regarding the absence of a 'reasonable and negotiated finite span of time' in which the Nuclear Weapons States (NWS) were to eliminate their nuclear weapons which made the treaty unequal and discriminatory. The treaty was termed at best a partial test ban treaty which Ghose stipulated to be not just flawed but also dangerous. As it banned only explosive testing (since most of the NWS had already conducted their explosions and were in a position to enhance their status by learning through extensive testing programmes, through more sophisticated and non-explosive technologies), she argued that the treaty would encourage a nuclear weapons technology race instead of an effective ban. India had supported a treaty which was multilaterally negotiated and universal, but its concerns were ignored and the text failed in the overarching objective of 'enhancement of international peace and security' mentioned in the mandate reiterated in the resolution 50/65 in 1995. All these led to the final decision as stated by Ghose, that 'India will never sign this unequal Treaty, not now, nor later'.

Ghose wrote 'Negotiating the CTBT' (Ghose, 1997) in the post-CTBT negotiation years. It provides a detailed explanation for India's position with regard to the treaty,

25 Statement in explanation of vote by Ms Arundhati Ghose; UN Archives.

which was based on the country's approach towards nuclear disarmament, its perception of nuclear weapons as a threat, its strategic conditions, as well as unanimous rejection of the 'unequal' treaty by the Indian Parliament. Ghose explained the importance of historical context in India's ethical approach in its foreign policy in general and nuclear issues in particular and stated that:

> While a country's position in arms control and disarmament negotiations is necessarily a product of its political, economic and strategic environment and its national security perceptions, it is equally a product of its unique historical experiences that have determined its fundamental world view.
>
> (Ibid: 250)

Ghose also traced India's experiences with the PTBT (1963) and the NPT (1968) and insisted that India achieving nuclear capability had more to do with its understanding of national security and strategic concerns than with issues of prestige or status. She emphasised the consistent and sincere commitment on India's part to complete global disarmament. According to Ghose, the GOI's Action Plan, 1988, was based on the premise that elimination of all nuclear weapons, and not joining the nuclear club, was in India's security interests.

Her account of the CTBT negotiations focused on the seamless attempts made by India and the other non-aligned countries to include a finite time frame for the process of disarmament and the rejection and ignoring of all such attempts at the conference which led to a lack of consensus. Moreover, it is argued that, by the end, attempts were made to corner India into signing the Treaty in its current form, which amounted to violation of the 1969 Vienna Convention on Law of Treaties.[26] Ghose also emphasized the assertion that India's position at the Conference was not a break but rather a continuance of India's approach on nuclear weapons.

Ghose also wrote extensively on bilateral and multilateral dealings regarding nuclear issues which had implications for India and the CTBT. In context of the 123 Agreement between India and the US in 2005 (the India–United States Civil Nuclear Agreement), Ghose emphasized the immense political influence over nuclear matters in the country (2009). The article portrays a sense of disapproval of the criticism posed to the Indo–US deal and the feasibility of signing the CTBT. She stated, 'there is always an option of discussing the technical issues after joining the treaty'. It need not necessarily be a condition for signature. Regarding the issue whether it would be in India's interest to sign the CTBT today, Ghose seemed more optimistic considering India's nuclear doctrine contains a 'no first use' clause, thus indicating an evolving position on the matter. It is evident that Ghose never shied away from expressing her informed opinion on important matters – whether it went with the government's decision, or in its criticism.

Another essay published by her in 2009 focused mainly

26 The Convention states that no state can be coerced into signing a treaty, nor can a treaty's entry into force be made conditional on the signature of any country.

on the specific aspects of the Fissile Material Cut-Off Treaty (FMCT), reasons for India's support to the FMCT, and strong arguments in favour of India signing the same. There is also an identifiable assertion of the FMCT being essentially and purely a non-proliferation treaty and not a disarmament treaty. Ghose's reference to the country's financial crisis and security imperatives reiterate her earlier position that decisions regarding nuclear issues in particular and important Indian national issues in general are ultimately political in nature. Ghose found India to be supportive, but reluctant at the same time. She pointed out that signing the FMCT is prudent for India (Ibid: 168) as since its nuclear doctrine is that of credible minimum deterrence, its stock of weapons grade fissile material is sufficient (in general and also in relative capacities to Pakistan and China).

Talking about the mutual commitment of Japan and India towards the idea of 'Nuclear Zero' in her lecture at a conference on the India-Japan strategic dialogue (2014), she highlighted that the countries' efforts have hardly been in cooperation with each other. Not only this, she reiterated her rejection of prestige, status, and domestic pressures as reasons for nuclear capability and emphasized that for a majority of states, the reason lies in security concerns.

She also provided evidence to corroborate that the Indian support for Nuclear Zero is not merely rhetorical in the conventional sense of the term. While discussing the credibility of the idea of a Nuclear Zero, Ghose asserted that nuclear weapons are already becoming obsolete with growing technology for and introduction of smaller, smarter and more efficient conventional weapons. Therefore, it would be sufficient if the weapons were delegitimized and their salience in the strategic doctrines of states reduced (Ibid: 44). Moreover, she argued for the use of rhetoric as the art of effective persuasion for those who look at nuclear weapons as the only way in which to protect their security.

Pakistan also features frequently in Ghose's academic writing. Arundhati Ghose and Manpreet Sethi together wrote about the stark differences between India and Pakistan[27] (Ghose and Sethi, 2015) as a reaction to Mark Fitzpatrick's suggestion about offering a nuclear deal to Pakistan – akin to that offered to India – in order to 'normalise' relations. They were concerned about the growing Pakistan narrative of a credible case for a deal, similar to the one offered to India, and argued that Pakistan was in fact facing serious challenges of terrorism, economic hardships and instability. Its proliferation history and brinkmanship behaviour further diluted its claim to participate in international nuclear commerce. By highlighting these differences, Ghose and Sethi made a serious attempt to break the western perception of treating both India and Pakistan as equals. The text shows an unmistakably sceptical approach towards Pakistan. For instance, they write that since there is a large section of uneducated youth in Pakistan, it presents itself as a rich pool for terrorist recruitment, the only industry that

27 Can Pakistan's Nuclear Dangers be 'Normalised'? *Survival – Global Politics and Strategy*, 2015.

could be said to be thriving in Pakistan (Ibid: 126).

A similar attitude towards the development of short range tactical weapons (Nasr) by Pakistan is seen in another publication by Ghose. In the article (Ghose, 2013), the author brings out the hypocritical standards of the western discourse with regard to the steps that India should take in order to deal with Pakistan in its increasingly overt proliferation and expansion of nuclear arsenal. She found the western discourse recommending India to observe restraint unacceptable as the Indian doctrine makes no distinction in categorisation of nuclear weapons, and any nuclear attack on Indian territory or on Indian troops anywhere else would be responded to by inflicting 'unacceptable damage'. Ghose argued that the appropriate manner to address the need for 'stability' in the region would involve addressing the questions to Pakistan. Secondly, Ghose argued that the Indian government should be more vocal and expressive about its reactions to recent developments affecting its nuclear posture. According to her, there is a need for an indication of policy changes to meet current challenges.

Apart from issues related to nuclear weapons, Ghose also worked on the WANA (West Asia North Africa) region (2015). In the context of the rapid political and social developments in WANA, Ghose highlighted the importance of the region for India, both as a source for energy imports, and as a host to a large number of Indian nationals, workers, and professionals. She therefore recommended that India should seek and support stability in the various states in the region. She therefore called for larger cooperation with the region which was undergoing a paradigm shift in ideologies and governance styles as the general population had become more aware of the power of street protests and the destabilising influences of political Islam (Ibid: 105). Her larger recommendation was for India to have increased interaction with the region through sustained dialogue along with an improvement in political, economic and commercial relations.

Ghose also made general observations and comments about the tactics used by various states in the region to deal with protests. Regions which did not use violence to curb the protests in 2011 and made certain economic concessions have a higher degree of stability than those which did not. Moreover, she highlighted the detrimental effect of extensive external intervention (from both regional and extra-regional powers) in the states of Libya and Syria.

Another sphere which saw significant academic contribution by Arundhati Ghose is that of Human Rights and the United Nations. Based on her experiences as India's Ambassador and Permanent Representative to the UN Offices in Geneva, Ghose made three specific remarks about the state of affairs in her work (2008). She highlighted the existence of high degree of politicisation of the subject with the North-South divide being evident in discussions, where countries of the North make accusations/charges against violations of human rights by states of the South. The second issue dealt with the focus on the source of violations – on violation of human rights by the state and its agents. The role of non-state actors like militant groups is rarely acknowledged. And finally, she emphasized the important role played by NGOs in the field of human rights. However, she reminded that even though there are organisations genuinely working for human rights, most of them are funded, directly or indirectly, by governments to project their own government's view. And interestingly, there is a sense that the NGOs have more credibility compared to the states from the South.

Arundhati Ghose was awarded the Friends of Liberation War Honour by the government of Bangladesh on 27 March 2012 in acknowledgment of her instrumental role in supporting the Bangladesh government in exile in 1971 during her three years of service in the newly formed state. Bhaskar Ghose describes a bond of trust, respect and friendship between the women of *Mukti Bahini*[28] and Arundhati Ghose, whom they lovingly called *didi* (elder sister).

Ghose remained active and engaged in professional and academic work throughout her life. This she did with immense sincerity and an impeccable intellectual honesty. She was a member and Chairman of UN Secretary General's Disarmament Advisory Board until 2001, a member of the Union Public Service Commission till 2004 and a member of UN Committee on Economic and Cultural Rights till 2005. Also, she was a member of the Executive Council of IDSA

28 The Mukti Bahini (Liberation Army) was the guerrilla resistance movement consisting of the Bangladeshi military, paramilitary, and civilians during the 1971 Bangladesh war for independence.

(now called Manohar Parrikar Institute for Defence Studies and Analyses) – a research organisation based in Delhi – from 2004 to 2007, and functioned for two years as Chairman of its Programme Committee. The period saw her active participation and guidance in the Track II diplomacy circuit. She was also a member of the Task Force on non-proliferation and disarmament set up by Ministry of External Affairs in 2007.[29]

Manpreet Sethi, a leading nuclear expert, remembers Arundhati Ghose as a great Ambassador for India, and a very supportive woman who was very happy to see another woman entering the nuclear issue field. While she does not recall Ghose overtly mention the fact of being a woman in a largely male-dominated field, the realization emerged in subtle undertones in her interactions. Sethi also highlights Ghose's support and guidance to young minds, and that she would often engage in insightful discussions with them about important issues at hand.

Arundhati Ghose was also a Distinguished Fellow at the Institute for Peace and Conflict Studies, Delhi, Global India Foundation, Kolkata and United Services Institution, Delhi. She was also Adjunct Fellow at the National Institute of Advanced Studies, Bangalore. She was on the editorial board of *Faultlines*[30], New Delhi and *India's National Security Annual Review*[31]. Ghose wrote extensively in journals, newspapers, and

29 CV from the website of National Institute of Advanced Studies.

30 *Periodical Journal*, magazine from the Institute for Conflict Management, New Delhi, India.

31 *Annual Review*, published under the aegis of Foundation for National Security Research, New Delhi, India.

books on disarmament and arms control issues, particularly nuclear issues, and was a regular speaker at various forums on these and related themes.

Conclusion

The wide range of work by the Ambassador indicates an evident inclination towards a realist perspective in her approach and writing. There is an almost unwavering and unflinching focus on the concept of 'national interest' and security understood in realist terms. She also finds (Ghose, 2014:43) James Acton and Perkovich's assessment, which identifies prestige, status and domestic pressures as possible motivating factors for states to acquire nuclear weapons, erroneous. For her, the reasons lie strictly in the realm of strategic concerns. However, she does not deny the role that unique historical experiences have played in the determination of India's fundamental world view as well as approach towards nuclear weapons (Ghose, 1997). Moreover, one finds Ghose consistently crediting the final say in nuclear matters to the political decisions of the day, thus indicating the significant importance of domestic political dynamics.

In Arundhati Ghose, India found an exemplary diplomat, a security expert, a vociferous negotiator and a true patriot. Her work was not limited to issues of security interests, and included engagement with issues of Human Rights, as well as other regions, including the WANA.

As concluding remarks, following is an excerpt from an interview of a Civil Services aspirant, with Arundhati Ghose as the Chairperson of the UPSC panel.[32] The excerpt is left as is, with the hope that it is self-explanatory, and provides a small insight into the person that was Arundhati Ghose:

> AG: Now, suppose you are in an Indian Embassy in some capacity below the rank of Ambassador. You get a direct order from Delhi, which you think is contrary to our national interests. Would you still carry it out?
>
> VK: As you said ma'am, I probably would have to.
>
> AG: (adopts a very nice tone) No! You must never do that. You should try and talk to the Ambassador and get him on your side. Once that is done, you can negotiate with New Delhi.
>
> VK: But ma'am, I was assuming that all this had been tried before. The order cannot come out of the blue. The matter must have been under consideration for some time and I would have discussed it with the Ambassador already.
>
> AG: Even then, you must not give up. The only guiding principle for you would be India's national interest. Nothing else. If you think something is wrong, refuse to carry out the order. The national interest must never be compromised.

REFERENCES

- Interview with Bhaskar Ghose, 20 August 2019.
- Interview with Manpreet Sethi, 29 January 2019.
- Telephonic interview with Sagarika Ghose, 3 August 2019.

32 Civil Services Exam Resource Centre 2007; Interview of Vivek Kumar

- Statement in explanation of vote by Ms Arundhati Ghose; UN Archives, URL:http://fas.org/news/india/1996/ctbt_UN_september_10_96.htm.
- Ghose, A. 'Negotiating the CTBT: India's Security Concerns and Nuclear Disarmament', *Journal of International Affairs*, Vol. 51(1), 1997.
- Ghose, A. 'India and the Fissile Material Cut-off Treaty', in Arvind Gupta (ed.) *India in a Changing Global Nuclear Order*, Academic Foundation, New Delhi p.159–168, 2009.
- Ghose, A. 'CTBT Redux', Institute of Peace and Conflict Studies, articles, #2840, 30 March, 2009.
- Ghose, A. 'The Road to Nuclear Zero: Rhetoric or Reality?', *CLAWS Journal*, Summer 2014, p. 36–45, 2014.
- Ghose, A. and M. Sethi, 'Can Pakistan's Nuclear Dangers be "Normalised"', *Survival – Global Politics and Strategy*, Feb-March, Vol. 57 (1): 117–131, 2015.
- Ghose, A. 'Western Discourse on Pak Nukes Exposes "Double Standards" Vis-à-Vis India', Vivekananda International Foundation, 28 Nov 2013, URL: http://www.vifindia.org/article/2013/november/28/western-discourse-on-pak-nukes-exposes-double-standards-vis-a-vis-india, 2013.
- Ghose, A. 'Transition in the Maghreb and Mashreq', in *West Asia in Transition*, Arundhati Ghose (ed.), IDSA & DPG, New Delhi, p. 105–112, 2015.
- Bhaskar, C.U. 'Arundhati Ghose (1940 – 2016): India's CTBT Durga', *The Indian Express*, 28 July 2016, New Delhi, 2016.

Arundhati Ghose

CONCLUSION
India on the Nuclear Map – an Assessment

Soumya Awasthi

And... '*The Buddha Smiles*'- Dr Raja Ramanna, former Director at Bhabha Atomic Research Centre (BARC) – 18 May 1974, 8:05 am.

India had long foreseen the need for atomic energy, which was not limited to military use, but had more significance for peaceful utilisation. More so, Dr Homi Bhabha, often called as 'Indian Oppenheimer' or the architect of the Indian Nuclear Programme, once stated, 'when nuclear energy has been successfully applied for power production, in say, a couple of decades from now, India will not have to look abroad for its experts but will find them ready at hand'. With statements such as this, Dr Bhabha emphasised on India being self-reliant in the field of nuclear science.

Today India is among the seven or eight countries in the world, and the only developing country, to have a complete fuel cycle, right from uranium explorations, mining, extraction and conversion through fuel lubrication and heavy water production reactors to reprocessing and waste management.

However, India did not have a smooth run-up to becoming a nuclear power. Since the Vedic period, India has been the land of inventors and thinkers producing outstanding and path-breaking research in various fields. Continuing with the tradition, India has produced some stellar scientists like Sir CV Raman, SN Bose, Sir JC Bose, Homi Bhabha and Homi Sethna, who have paved the way for organised research in nuclear science. Indian Parliament passed the Atomic Energy Act in 1948 and set forth India's objective of developing atomic energy for peaceful purposes. After a few years of struggle, the government of independent India established the Department of Atomic Energy in 1954, founded by Homi J Bhabha, with its headquarters in Mumbai, with the sole task of exploring the nuclear field in the country. Since then, DAE has developed nuclear and radiation technology in agriculture, medicine, industry and basic research.

A common predicament of developing countries was that technologies imported from the developed countries were challenging to adapt and implement as per the socio-economic and local conditions. However, India laid the groundwork for an intricate nuclear science programme in 1948, soon after independence. After a decade of atomic contradiction, India's nuclear weapons program became the most talked about and debated nuclear weapons program in the history of all nuclear states. As George Perkovich writes, 'no nation debated more democratically than India on whether to get or give up nuclear power guns' (Perkovich, 2002).

For the first four decades, India battled with the idea of building or not building a nuclear bomb. There were several negotiations and debates between Bhabha, Sarabhai and the government of India. After a long period of laxity, India decided to step into the new phase with Prime Minister Shastri allowing nuclear scientists to study the underground nuclear explosion process (Chengappa, 2002). It is worth noting that India never wanted to carry out an explosion, so it asked for nuclear security from nuclear powers against the Chinese. However, these nuclear powers did not give any affirmative response to India. As a result, the government of India under Indira Gandhi declined to sign the Nuclear Non-Proliferation Treaty (NPT). It was after the 1971 Bangladesh Liberation war that India, under the leadership of Prime Minister Indira Gandhi, gave the green signal to conduct a peaceful nuclear test which was weaponized only in 1989 under Prime Minister Rajiv Gandhi (Subrahmanyam, 1998). The decision to weaponize the Nuclear test was taken under vulnerable circumstances, with Pakistan testing a nuclear weapon with the help of China and the United States (Subrahmanyam, 1998).

After decades of ambiguities, India decided to be a responsible nation, and under the leadership of Prime Minister Vajpayee, India raised the curtain of secrecy and pursued the nuclear programme. But this did not mean that India would amass a sizeable nuclear arsenal. Even though India had all the capability to produce a large quantity of arsenal, it decided to restrain itself and exhibit responsible behaviour.

It adopted the deterrent-only nuclear use doctrine. As former Foreign Secretary Shyam Saran says, India chose responsibility and restraint (Saran, 2014) to pursue a 'minimum credible deterrent' (Chari, 2000). Since the 1998 tests, India's nuclear arsenal has seen a gradual increase from maybe a dozen in 1998 to approximately 150 warheads. What else can be a better example of India's responsible and restrained behaviour than the Kargil War in 1999 when despite continuous requests from the Chief of Staff and the National Security Advisory Board (NSAB), the Vajpayee government did not permit the use of a nuclear weapon and India paid the exorbitant human cost.

Ultimately, the 2008 Indo–US nuclear deal pushed India into the path of reviving nuclear power. This phase was driven mainly by India's rising security concerns vis-à-vis China and Pakistan. India's nuclear weapons project is focused on its security interests. India's objective is to achieve deterrence against China over a period. This has gained traction, especially now, due to the boundary disputes and increasing military mobilisation. With the Indo–US nuclear deal, India has now been accommodated into major technology denial regimes like the Australia Group, the Wassenaar Arrangement and the Missile Technology Control Regime, making India a critical technology partner of the Western countries.

However, despite India being accommodated at various platforms, its responsible attitude has continued to be conservative in producing nuclear warheads, though it has produced many delivery vehicles and Intermediate-Range Ballistic Missiles (Kristensen and Korda, 2020). Apart from land to air, India also advanced itself in the sea to air missile programme. Agni II, Agni III, Agni IV, Agni V, and Ballistic Nuclear Submarine (SSBN) have increased their lethality and accuracy. India is slowly moving closer to a launch-on-command nuclear posture from a de-alerted and de-mated nuclear arsenal (Joshi, 2022). Beyond the advanced nuclear capabilities, India is deliberating over its doctrine which got a push after the Kargil War in 1999 when India decided on the No-First-Use (NFU) policy. India never intended to use nuclear weapons for offensive measures; they were simply for deterrence. India's nuclear strategy is based on credible minimum deterrence with second strike capability. However, in 2019, Defence Minister Rajnath Singh, on the occasion of the death anniversary of former Prime Minister Atal Bihari Vajpayee, made a statement that India's NFU is conditional and can change (*The Hindu*, 2019). From time to time, the debate on NFU has heated up and faced criticism from strategists arguing that due to NFU, often dealing with adversary Pakistan becomes tricky and limits the options.

India's resurgent attitude was also visible when it decided not to become a signatory of the Comprehensive Test Ban Treaty (CTBT) and Non-Proliferation Treaty (NPT). It has been half a century since India refused to sign the Non-Proliferation Treaty, which was written to save the world from a catastrophic event. NPT was considered the cornerstone of the global non-proliferation regime, but it failed to ensure disarmament. Although India played a very significant role in negotiating the NPT, it refused to sign the treaty because of its biased nature against the national interests of India. The treaty had two distinct safeguard standards for parties – a higher standard for the 'have-nots' and a lower standard for the 'haves', which proved to be feckless to the treaty's objective. India suggested that step-by-step nuclear reduction approach was a way to go about it instead of imposing the decisions on the developing countries. India then chose to conduct a peaceful nuclear explosion in 1974. India argued that it was an essential feature of national security when surrounded by nuclear states like China and Pakistan. And after almost two decades of pause, on 11 May 1998, India, under the leadership of the Prime Minister, carried out three tests to conclude its underground tests. Defence Minister George Fernandes labelled China as its first threat. Home Minister LK Advani claimed Pakistan to be another threat that needed to be dealt with firmly (Tellis, 2001).

This invited heavy economic sanctions from the United States, which was rooting for signing the Comprehensive Test Ban Treaty (CTBT). India, however, remained supportive of the Fissile Material Cut-off Treaty (FMCT). Thus, India shielded its freedom of action concerning its nuclear status. Consequently, India can produce fissile materials like plutonium (Brian, Speier and Jones, 1995). In his book *India's Emerging Nuclear Posture (2001),* Ashley Tellis argues that four variables have influenced India's nuclear posture and will influence it in times. These are, first, the nature of the global nuclear regime, second, demand imposed by regional threats and security, third, India's relations with key global powers and international organisations, and fourth would be, opportunities it has from the nuclearisation debates in political circles of India (Tellis, 2001).

Recent events in the northern sector of India like Pulwama-Balakot in 2019 and Galwan in 2020 justify why India still needs to maintain sufficient deterrence due to multifaceted security challenges. Therefore, though the current global nuclear order points towards denuclearisation, it is impossible at this juncture.

India has been committed to the NFU policy as far as its nuclear warfare strategy is concerned. It has been a significant feature of India's nuclear posture, which has made it a responsible state. As a result, it continued to earn silent support from the global powers. However, India has never advocated going 'nuclear' in times of war; instead it has always talked about the peaceful use of nuclear energy. K Subrahmanyam is one of the prominent personalities who helped India adopt an NFU posture.

In 2008, the Government of India offered to assist the United Nations in eliminating nuclear weapons in phases. In the first phase, all countries should negotiate a global no-first-use agreement; secondly, all countries should negotiate non-use of nuclear weapons against non-nuclear-weapon states; thirdly, all countries should reduce the dependence on nuclear weapons in their security doctrine and finally, the establishment of global non-discriminatory verifiable time-bound elimination.

However, since the abrogation of Article 370 in Jammu and Kashmir, there has been a debate over Defence Minister Rajnath Singh's statement that India can explore the nuclear option if the need arises. Simultaneously, several Indian strategists and policymakers argued that India should revisit NFU. The world is now entering a 'Second Nuclear Age' (Paul Bracken, 2012), which is influenced by information technology, artificial intelligence, cloud computing, data analytics, cyber warfare and above all, the threat of terrorists hijacking the nuclear weapons. This has caused the nuclear age to shift from Europe to Asia. Technological advancement and rapidly evolving global diplomacy have influenced the shaping of strategic culture around the globe. Moreover, the current trends of nuclearisation in south Asia confirm that countries in Asia have been blinded by obduracy to build more and more weapons in the spirit of international competition.

China challenging America with every available resource for global dominance, has made nuclear weapons an essential instrument of security calculus at global and regional levels. China also faces rivals like Japan, Australia, Vietnam, Taiwan and the Philippines. India's Nuclear Doctrine never envisioned nuclear weapons as the security structure's centrepiece. But India has its justifications for holding on to the freedom to build nuclear weapons as it is wedged between two nuclear rivals, Pakistan and China. They have shown hostility towards India on several occasions, including full-scale wars in 1962, 1965, and 1999. Today, India has witnessed transformations in China's foreign policy, and Pakistan-sponsored terrorism has forced Indian elites to deliberate on some features of the doctrine. As a result, these dynamics have become the underlying forces leading to deliberations on the NFU policy.

As Pandit Nehru said in 1948, while speaking to the

members of the Constituent Assembly,

> … of course, if we are compelled to use [atomic energy] for another purpose, India will not restrain from using it that way. But I pray that we are not forced to that extent, and atomic energy is going to be a peaceful one…and not one of war and hatred.

Ultimately, due to India's continued honouring of the credible minimum deterrence, the production of warheads has not dramatically escalated since the 1998 tests, despite unofficial debates (Tellis, 2022).

To conclude, India's journey to nuclearisation has been extended, most debated and calculated, unlike the Chinese or any other country's nuclear programme. However, there is a need for a more nuanced debate on the subject, keeping in mind the long-term impact of present scenarios.

India's journey to find its place on the map of nuclear power nations was no less than a Mozart symphony with several major and minor chords orchestrated by the talented scientists, strategists, diplomatic and political communities of India.

MILESTONES IN INDIA'S NUCLEAR JOURNEY

1944-45: Dr Homi Jehangir Bhabha established a nuclear research centre called the Tata Institute of Fundamental Research in Mumbai.

1948: The Atomic Energy Commission (AEC) was established under the direction of Dr Homi J. Bhabha.

1950: Indian Rare Earths Limited (IREL) was established. It was a joint venture between the Government of India and the Government of Travancore.

1951: The first uranium deposit in India was discovered in Jaduguda, Jharkhand.

1954: The Department of Atomic Energy (DAE) was created.

1954: India proposed a Standstill Agreement. Pandit Nehru was the first global leader to have called for an end to all nuclear testing. In his words, he rejected 'atomic colonialism'.

1955: The Asian-African Conference at Bandung, Indonesia, called for 'universal disarmament and effective international control to ensure it'.

1956: India's one-megawatt Apsara Research Reactor attained criticality.

1957: India established the Atomic Energy Establishment, Trombay, which was renamed the Bhabha Atomic Research Centre (BARC) in 1967.

1959: The Uranium Metal Plant at Trombay began production.

1962: Chinese attack on India impeded the development of nuclear weapons.

1962: Heavy water production began at the German-built Nangal plant in Punjab. It is one of the oldest heavy-water plants.

1964: China conducted its first nuclear test in October 1964. The test sparked demands that India must develop a deterrent.

1964: The Subterranean Nuclear Explosions Project (SNEP) was initiated. The then Prime Minister Lal Bahadur Shastri allowed the scientific nuclear establishment led by Dr Bhabha to research peaceful uses of nuclear energy – the Study of Nuclear Explosions for Peaceful Purposes (SNEPP).

1967: Nuclear Programme of India starts.

1968: India refused to sign Nuclear Non-Proliferation Treaty (NPT).

1968: Nuclear Fuel Complex was established at Hyderabad under the DAE.

1974: Operation 'Smiling Buddha' or Pokhran-I nuclear test was conducted. India carried out an underground nuclear explosion at Pokhran, Rajasthan. It was a Peaceful Nuclear Explosion (PNE).

1982: India launched the Integrated Guided Missile Development

Programme (IGMDP) to research and develop a comprehensive range of Missiles. It included missiles like Prithvi, Agni, Akash, Trishul, and Nag.

1988: India proposed Rajiv Gandhi Action Plan on Disarmament for a nuclear-weapon-free world. However, the action plan failed to convince the Nuclear Weapon States; instead, Pakistan acquired a bomb too.

1988: Pakistan and India agreed to exchange lists of nuclear installations as part of an agreement not to attack each other's nuclear facilities. The first exchange occurred in January 1992.

1988: The INS Chakra nuclear submarine was leased from Russia for three years and arrived in Visakhapatnam, India.

1989: DRDO developed a short-range ballistic missile AGNI-I and tested it at Chandipur, Odisha, India.

1990: US President George Bush eased export restrictions on supercomputer exports to Brazil, India, and China.

1991: The Indian Navy began working on a nuclear-powered submarine project shortly after returning a Charlie I-type SSGN leased from the Soviet Union.

1995: India inaugurated the Narwapahar uranium mine in Jharkhand.

1995: India refused to sign the Comprehensive Test Ban Treaty (CTBT). Ambassador, Arundhati Ghose, the Permanent Representative of India to the UN at Geneva during the CTBT negotiations, argued that while India was looking to cap vertical proliferation, the NWS were concerned with horizontal expansion. Two other developments sealed India's decision: the cementing of the China–Pakistan missile and nuclear cooperation and the resumption of American military aid to Pakistan. Ultimately on 20 June 1996, India rejected the treaty with Ambassador Ghose, invoking 'national security considerations.'

1997: Prime Minister IK Gujral says India will not sign the Fissile Material Cut-off Treaty (FMCT) or any other 'discriminatory' nuclear agreement that would hamper India's nuclear program.

1998: Pokhran-II, a series of five nuclear tests was conducted, including the first fusion bomb. India conducted two rounds of nuclear weapon tests. After the first round, Prime Minister Atal Bihari Vajpayee announced that 'a fission device, a low-yield device, and a thermonuclear device' had been successfully tested in the deserts of Pokhran. The government exploded two more sub-kiloton nuclear tests at the same testing range two days later. The five underground tests ranged in yield from less than one kiloton to an estimated 45 kilotons.

1998: President Clinton imposed economic sanctions on India after it refused American demands to disavow future testing or deployment of nuclear weapons.

1998: Russia refused to join other countries in punishing India for its nuclear tests.

1998: In response to India's nuclear tests, the World Bank postponed the approval of $865 million in loans to India.

1998: The US Department of Commerce's Bureau of Export Administration imposed sanctions on the government and private entities which were thought to be involved in nuclear or missile activities.

1998: Indian Prime Minister Atal Bihari Vajpayee told the parliament that India's nuclear doctrine will be centred on a small but credible deterrent and a no-first-use policy.

1999: Announcement of India's Nuclear Doctrine and NFU.

1999: The Assistant Secretary for Export Administration at the US Department of Commerce announced the removal of 51 organisations from the list of 200 Indian entities sanctioned in November 1998.

2001: The Canadian government announced lifting economic sanctions imposed on India after its May 1998 nuclear tests.

2001: US President George W Bush waived US economic sanctions against India initially imposed as a penalty for the nuclear weapon tests conducted in 1998.

2001: Japan lifted the economic sanctions imposed on India after the May 1998 nuclear weapon tests.

2001: India and the United States resumed military-to-military cooperation and revived the Defense Policy Group (DPG), which was suspended after India's May 1998 nuclear tests.

2002: The Chairman of the AEC in India, Dr Anil Kakodkar, unveiled a Rs100-crore program that focused on the use of thorium as an alternative to uranium in nuclear energy generation.

2003: Establishment of Nuclear Command Authority.

2003: India established its Strategic Forces Command (SFC) and approved the appointment of a Commander-in-Chief to manage its nuclear and strategic forces.

2003: India outlined its eight-point nuclear doctrine. The doctrine included among others 1) a no-first-use posture, 2) building and maintaining a credible minimum deterrent, and 3) the option to use nuclear weapons in retaliation to chemical and biological attacks.

2005: India passed the Weapons of Mass Destruction and their Delivery Systems Bill in response to UN Security Council Resolution 1540.

2006: DAE Secretary Anil Kakodkar recommended that, besides the Dhruva and CIRUS units, a portion of India's nuclear reactors, including the breeders and the naval reactor, not be put under IAEA safeguards to meet the country's 'strategic need'.

2006: US President George W Bush signed the United States-India Peaceful Atomic Energy Cooperation Act, an essential step in enabling the United States to share civilian nuclear technology with India.

2007: India and Pakistan signed an agreement on 'Reducing the Risk from Accidents Relating to Nuclear Weapons' that required both countries to immediately notify each other of any nuclear-weapon-related accident that could create cross-border radioactive fallout risk or an outbreak of nuclear war. The agreement was renewed for five years, in 2012 and 2017.

2008: India signed a civil nuclear cooperation agreement with the US, allowing India to sell and purchase nuclear fuel and technology from the United States.

2009: India launched the INS Arihant, its first nuclear-powered submarine, which will reportedly be capable of launching nuclear-capable ballistic weapons.

2011: India undertook seven successful tests of AGNI-IV between 2011-2022. The last one in June 2022 was an Intermediate Range

Ballistic Missile (IRBM) training launch in APJ Abdul Kalam Island, Odisha. The missile is assessed to have a 4000 km range capability.

2012: The Indian Navy inducted a second INS Chakra nuclear submarine leased from Russia into service.

2012: India successfully conducted a launch test of AGNI-V off the coast of Odisha. The first test launch was in 2012, and then there were seven consecutive launches. The last one, in December 2022, was the first night trial of AGNI-V. The launch also proved striking capability beyond 7000 km.

2014: The INS Arihant nuclear submarine reportedly began sea trials.

2017: The INS Arighat nuclear submarine (originally called INS Aridhaman) was quietly launched. The Arighat is believed to be able to carry more SLBMs than the Arihant and to have a more powerful reactor.

2017: Indian Navy Chief Admiral Sunil Lanba reportedly confirmed that India has started to build six nuclear-powered attack submarines.

Sources:

- Jonathan McLaughlin, 'India Nuclear Milestones: 1945-2018', 31 October 2018, Wisconsin Project on Nuclear Arms Control.
- Sameer Patil and Arun Vishwanathan, 'India and Nuclear Disarmament: A Quest Rooted in National Security Considerations', *India Quarterly* 78(2) 261–279, 2022.

ABBREVIATIONS

AEC: Atomic Energy Commission
AEE: Atomic Energy Establishment
AERC: Atomic Energy Research Commission
AERE: Atomic Energy Research Establishment
AICC: All India Congress Committee
AIIMS: All India Institute of Medical Sciences
AIYC: All India Youth Congress
APPLE: Ariane Passenger Payload Experiment
ATIRA: Ahmedabad Textile Industry's Research Association
AVM: Air Vice Marshall
AVSM: Ati Vishisht Seva Medal
BARC: Bhabha Atomic Research Centre
BJP: Bharatiya Janata Party
BJS: Bharatiya Jana Sangh
BRI: Belt and Road Initiative
CANDU: Canada Deuterium Uranium
CAPS: Centre for Air Power Studies

CIA: Central Intelligence Agency

CIRUS: Canada India Reactor Utility Services

Cmde: Commodore

COD: Conference on Disarmament

COAS: Chief of Army Staff

CPI: Communist Party of India

CTBT: Comprehensive Nuclear-Test-Ban Treaty

CWC: Congress Working Committee

DAE: Department of Atomic Energy

DDA: Delhi Development Authority

DRDL: Defence Research and Development Laboratory

DRDO: Defence Research and Development Organisation

EC: European Commission

ELV: Expendable Launch Vehicle

FBTR: Fast Breeder Test Reactor

GFRP: Glass Fiber Reinforced Polymer/Plastics

HRD: Human Resource Development

IAF: Indian Air Force

IAEA: International Atomic Energy Agency

IASc: Indian Academy of Sciences

IDSA: Institute for Defence Studies and Analyses

ICWA: Indian Council of World Affairs

ICSSR: Indian Council of Social Science Research

IIC: India International Centre

IIM-A: Indian Institute of Management, Ahmedabad

IISc: Indian Institute of Science

IGMDP: Integrated Guided Missile Development Programme

IMD: India Meteorological Department

INCOSPAR: Indian National Committee for Space Research

ISRO: Indian Space Research Organisation

LoC: Line of Control

MIT: Madras Institute of Technology

MP: Member of Parliament

NIAS: National Institute of Advanced Studies

NFU: No First Use

NDA: National Democratic Alliance

NGO: Non-Governmental Organisation

NPT: Nuclear Non-Proliferation Treaty

NSA: National Security Advisor

NSCAB: National Security Council Advisory Board

NSG: Nuclear Suppliers Group

NTI: National Training Institute

NNWS: Non-Nuclear Weapon State

NWS: Nuclear Weapon State

OPEC: Organization of Petroleum Exporting Countries

PINSTECH: Pakistan Institute of Nuclear Science and Technology

PM: Prime Minister

PMO: Prime Minister's Office

PNEs: Peaceful Nuclear Explosions

PRL: Physical Research Laboratory

PSLV: Polar Satellite Launch Vehicle

PTBT: Partial Test Ban Treaty

RAPS: Rajasthan Atomic Power Station

RATO: Rocket Assisted Take Off

RSS: Rashtriya Swayamsevak Sangh

SFC: Strategic Forces Command
SNEP: Subterranean Nuclear Explosion Project
SNEPP: Study of Nuclear Explosions for Peaceful Purposes
SSTC: Space Science and Technology Centre
STS: Semipalatinsk Test Site
SWDT: Special Weapons Development Team
TERLS: Thumba Equatorial Rocket Launching Station
TIFAC: Technology Information Forecasting and Assessment Council
TIFR: Tata Institute of Fundamental Research
UGC: University Grants Commission
UN: United Nations
UNDC: United Nations Disarmament Commission
UNESCO: United Nations Educational, Scientific and Cultural Organization
UNGA: United Nations General Assembly
US: United States
USSR: Union of Soviet Socialist Republics
VECC: Variable Energy Cyclotron Centre, Kolkata
VM: Vayu Sena Medal
VrC: Vir Chakra
VSSC: Vikram Sarabhai Space Centre
WANA: West Asia-North Africa

LIST OF CONTRIBUTORS

Rajarshree Narayan Chowdhury

Rajarshree Chowdhury has received a PhD from the Centre for International Politics, Organization and Disarmament (CIPOD) at Jawaharlal Nehru University (JNU). His present research interests include democracy and grassroots politics, public participation in local governance. He has presented ideas across national and international forums like the Graduate Research Meet at Indian Institute of Technology, Guwahati (IIT-GRM), British Association of South Asian Studies (BASAS) and the Political Studies Association (PSA). His research has recently been published in the *Journal of Asian and African Studies*.

Arun Vishwanathan

Arun Vishwanathan is currently Associate Professor in Security Studies in the School of National Security Studies, Central University of Gujarat, Gandhinagar. In the past, he has held the position of Assistant Director in the National Security Council Secretariat between 2008 and 2011 and was Associate Fellow at the Indian Pugwash Society, IDSA Campus, New Delhi between 2005-2008. Vishwanathan is a co-editor of the book *Troubling Tehran: Reflections on Geopolitics* and co-author of the monograph Hatf-IX/ *NASR – Pakistan's Tactical Nuclear Weapon: Implications for Indo–Pak Deterrence.* His research papers have been published in *Strategic Analysis, USI Journal, Contemporary Review of the Middle East, International Journal of South Asian Studies, Synergy: Journal of Centre for Joint Warfare Studies* in addition to opinion pieces in various national and international media houses.

Sameer Patil

Sameer Patil is a Senior Fellow at Observer Research Foundation (ORF), Mumbai. His work focuses on the intersection of technology and national security, including cybersecurity. He also researches India's national security priorities, counter-terrorism, and regional security.

Patil is the author of *Securing India in the Cyber Era* (Routledge, 2022) and has co-edited *Moving Forward EU-India Relations: The Significance of the Security Dialogues* (Edizioni Nuova Cultura, 2017). He has a PhD and an MPhil in international politics from Jawaharlal Nehru University, New Delhi. Patil has participated in several Track 1 and Track 1.5 dialogues, including India–US and India–UK Strategic Intelligence Dialogues, after the 2008 Mumbai attacks, and India–UK Track 1.5 Cyber Dialogue. In 2019, he received the prestigious Canberra Fellowship, awarded by the Department of Foreign Affairs and Trade, Australia. His work and views have been cited by *The New York Times, The Washington Post,*

The Guardian, Financial Times, The Wall Street Journal, Al Jazeera, CNBC, Associated Press, and other news agencies.

Shrabana Barua

Shrabana Barua is a Research Fellow at the Indian Council of World Affairs (ICWA), Sapru House, New Delhi. Prior to this she was an Assistant. She is also a Visiting Fellow at NIICE, Kathmandu. Her research focuses on geopolitics of South and Southeast Asia, particularly Pakistan, China, India and Myanmar. Her PhD is from Centre for International Politics, Organization and Disarmament (CIPOD), School of International Studies, Jawaharlal Nehru University, New Delhi. She holds a Bachelor's and Master's degree in Political Science from Hindu College, University of Delhi. Earlier, she worked with Gateway House, Mumbai. She has published several articles and book reviews in reputed journals and presented papers at national and international conferences. She writes regularly for the *Hindustan Times* on global politics.

Rajaram Nagappa

Rajaram Nagappa has specialised in aerospace propulsion. He has worked extensively in the design and development of solid propellant rocket systems. His interests are in missile technology and space weaponisation.

He has served in the Vikram Sarabhai Space Centre, ISRO as its Associate Director, and later was Pandalai Memorial Chair Professor at Anna University, Chennai. He has also taught at Technion – Israel Institute of Technology, Israel. He has received several awards including the Astronautical Society of India Award, Distinguished

Alumnus Award of the Madras Institute of Technology, DRDO's Agni Award for Excellence in Self Reliance, Certificate of Appreciation of the International Astronautical Federation and the Honorary Fellowship of the High Energy Materials Society of India. His recent works include an assessment of the Pakistani cruise missiles and an assessment of the Iranian satellite launch vehicle Safir. He has also traced the development of fighter aircraft in China as part of a study on China's Science & Technology capability.

Sitakanta Mishra

Sitakanta Mishra is an Associate Professor in the School of Liberal Studies at the Pandit Deendayal Petroleum University; Associate Member of Pugwash India, New Delhi; Honorary Member of the Editorial Board of the IJBST Journal Group; Member of International Studies Association (ISA); Member of Pacific Forum, Honolulu; Associate Member, Institute of Defence Studies & Analyses (IDSA), New Delhi; Non-Resident Research Fellow, Centre for Air Power Studies (CAPS), New Delhi; Life Member of United Service Institution of India (USI), New Delhi; Reviewer of *Journal of Management and Sustainability*, Canadian Center of Science and Education; Peer Reviewer of *Journal of Asian Security and International Affairs* (Sage); Member of Task Force to study India's Energy Security in a Carbon Constrained World by Vivekananda International Foundation, New Delhi.

Yasemin Develi

Yasemin Develi is a PhD researcher at King's College, London, working on Indian Foreign Policy. Her research seeks to explain the impact of Foreign Policy Executives, in particular Prime Ministers, on change and continuity in Indian Foreign Policy. To this end, she analyses the political personality and leadership of Atal Bihari Vajpayee and his interaction with key decision-makers, as well as the larger cabinet. This research is concerned with understanding the decision-making process under the leadership of Vajpayee and gauging the impact he had on this process.

Soumya Awasthi

Soumya Awasthi is a Consultant with the Tony Blair Foundation. Previously she was an Associate Fellow at the Vivekananda International Foundation (VIF), New Delhi. Earlier, she has worked with Manohar Parrikar Institute for Defence and Strategic Analyses (MP-IDSA), New Delhi. She has also worked as a Fellow at Centre for Armed Forces Historical Research (CAFHR) in United Service Institution, New Delhi and Centre for United Nations Peacekeeping Organization, New Delhi. Her research interests include Counter Terrorism, De-radicalization, South Asian strategic studies, India's defence and strategic policy, and Afghanistan. She is particularly interested in the Deoband School of Islam in India, Pakistan and Afghanistan. Her PhD is from the Centre for International Politics, Organization, and Disarmament (CIPOD) at School of International Studies (SIS) in Jawaharlal Nehru University on the subject 'Role of Religious Discourse in Regional and Global Geopolitics: Case Study of Deoband Philosophy in India, Pakistan and Afghanistan'. She has completed her MPhil on the topic 'Defence Budget of India from 1947 till 2015: A Trend Analysis' in the Department of Defence and Strategic Studies at the Savitribai Phule Pune University,

Maharashtra. Her work and views have been published in several national and international forums.

Manpreet Sethi

Manpreet Sethi is a Distinguished Fellow at Centre for Air Power Studies, New Delhi where she leads its nuclear project. Having worked on all nuclear issues – ranging from energy and strategy to non-proliferation and disarmament – for over two decades now, she was awarded the K Subrahmanyam Award (2014) for excellence in strategic and security studies. She received commendation by the Chief of the Air Staff (2019) for her professional ability and dedication to the cause of national security. She has over 130 articles in reputed publications and eight books to her name.

Renjie Feng

Renjie Feng's research focuses on geopolitics of South Asia, India-China Relations, climate change policies and global disarmament issues. He obtained his PhD and MPhil degrees in 2022 and 2017 respectively, from the Centre for International Politics, Organization and Disarmament (CIPOD), School of International Studies at JNU. He has published several articles and book reviews in reputed academic journals, including *China Report* (Sage), *East Asia Policy* (World Scientific), *China Quarterly of International Strategic Studies* (World Scientific), *International Studies* (Sage), etcetera. He worked as a research intern at the Institute of Chinese Studies, Delhi. He has presented papers at national and international conferences conducted by institutions such as the International Studies Association (ISA), University of Rajasthan, and Tsinghua University (Beijing). He has also published many commentaries on relevant issues on websites such as Rediff (India) and Guancha (China).

Vijaylaxmi Yadav

Vijaylaxmi is currently a Doctoral Candidate in Political Geography at the Centre for International Politics, Organization, and Disarmament, School of International Studies, JNU. She is working on 'Alternative Lives of Contemporary Travellers and Artists'. She has previously worked on 'Space and Sovereignty in Cyberspace'. Her fields of interest include interdisciplinary studies including critical theory, geopolitics, cultural history and social anthropology.